Semantic Interoperability: Issues, Solutions, and Challenges

RIVER PUBLISHERS SERIES IN INFORMATION SCIENCE AND TECHNOLOGY

Consulting Series Editor

KWANG-CHENG CHEN
National Taiwan University
Taiwan

Information science and technology enables 21st century into an Internet and multimedia era. Multimedia means the theory and application of filtering, coding, estimating, analyzing, detecting and recognizing, synthesizing, classifying, recording, and reproducing signals by digital and/or analog devices or techniques, while the scope of "signal" includes audio, video, speech, image, musical, multimedia, data/content, geophysical, sonar/radar, bio/medical, sensation, etc. Networking suggests transportation of such multimedia contents among nodes in communication and/or computer networks, to facilitate the ultimate Internet. Theory, technologies, protocols and standards, applications/ services, practice and implementation of wired/wireless networking are all within the scope of this series. We further extend the scope for 21st century life through the knowledge in robotics, machine learning, cognitive science, pattern recognition, quantum/biological/molecular computation and information processing, and applications to health and society advance.

- Communication/Computer Networking Technologies and Applications
- Queuing Theory, Optimization, Operation Research, Statistical Theory and Applications
- Multimedia/Speech/Video Processing, Theory and Applications of Signal Processing
- Computation and Information Processing, Machine Intelligence, Cognitive Science, and Decision

For a list of other books in this series, please visit www.riverpublishers.com

Semantic Interoperability: Issues, Solutions, and Challenges

Editors

Salvatore F. Pileggi
ITACA-TSB, Universidad Politécnica de Valencia, Spain

and

Carlos Fernandez-Llatas
ITACA-TSB, Universidad Politécnica de Valencia, Spain

LONDON AND NEW YORK

Published 2012 by River Publishers
River Publishers
Alsbjergvej 10, 9260 Gistrup, Denmark
www.riverpublishers.com

Distributed exclusively by Routledge
4 Park Square, Milton Park, Abingdon, Oxon OX14 4RN
605 Third Avenue, New York, NY 10158

First published in paperback 2024

Semantic Interoperability: Issues, Solutions, and Challenges / by Salvatore F. Pileggi, Carlos Fernandez-Llatas.

Routledge is an imprint of the Taylor & Francis Group, an informa business

Publisher's Note
The publisher has gone to great lengths to ensure the quality of this reprint but points out that some imperfections in the original copies may be apparent.

While every effort is made to provide dependable information, the publisher, authors, and editors cannot be held responsible for any errors or omissions.

ISBN: 978-87-92329-79-0 (hbk)
ISBN: 978-87-7004-527-8 (pbk)
ISBN: 978-1-003-33946-5 (ebk)

DOI: 10.1201/9781003339465

Table of Contents

PART I

Towards Semantic Interoperability

Introduction

Salvatore F. Pileggi and Carlos Fernandez-Llatas

ITACA-TSB Universidad Politécnica de Valencia, Camino de Vera S/N, 46022 Valencia, Spain; e-mail: {salpi, cfllatas}@itaca.upv.es

Towards Semantic Interoperability

In our current global society, there is a growing necessity in the inter-communication among different actors available in defined processes. The incorporation of computers as actors in those systems poses a problem in that field. Computers traditionally use syntactically defined data in order to communicate among themselves. With the arrival of the necessity of Human Computer interaction, the use of semantic information in the communication becomes crucial for an independent, effective information interchange among the actors, whether they are computers or human beings.

For this, semantic technologies are experiencing an increasing popularity in the context of different domains and applications. The idea to have a semantic language that allows communication unification among the actors will enable the system to deploy a better model of understanding. Using this idea, the way of communicating between different systems changes. The understanding of any class of system can be significantly changed under the assumption that any system is part of a global ecosystem known as the Semantic Web. The Semantic Web is an evolving extension of the current Web model (normally referred to as the Syntactic Web) that introduces a semantic layer in which semantics, or the meaning of information, is formally defined.

So, semantics should integrate web-centric standard information infrastructures improving several aspects of interaction among heterogeneous systems. This is because common interoperability models are progressively becoming obsolete if compared with the intrinsic complexity and always more distributed focus that modern systems feature. For example, the basic interoperability model, that assumes the interchange of messages among sys-

S.F. Pileggi and C. Fernandez-Llatas (Eds.), Semantic Interoperability: Issues, Solutions, and Challenges, 3–5.

tems without any interpretation, is simple, but effective only in the context of close environments. Also more advanced models, such as the functional interoperability model that integrates the basic interoperability model with the ability of intepretating data context under the assumption of a shared schema for data fields accessing appears unable to provide a full sustainable technologic support for open systems.

The Semantic Interoperability model would improve common interoperability models introducing the interpretation of the meaning of data. Semantic interoperability is a concretely applicable interaction model under the assumption of adopting rich data models (commonly called ontology) composed of concepts within a domain, and the relationships among those concepts.

In practice, semantic technologies partially invert the common view at actor intelligence: intelligence is not implemented (only) by actors but it is implicitly resident in the knowledge model. In other words, schemas contain information and the *code* to interprete it.

For all this, Semantic Interoperability is one of the most important research fields. The use of semantic technologies to communicate between systems not only allows a better way to share knowledge among the most heterogeneous systems, but also to provide a great added value to the applications designed in almost all fields in terms of Human Computer Interaction. In this way, the incorporation of semantics into computer languages can radically improve the relationship between humans and applications. Using that technology, the applications can gather richer knowledge about the user necessities in order to provide a better way to understand them and offer the information in a clearer, more individualized way. There are lots of examples of such use of semantics. With this technology, search engines will be able to gather more precise information for users, avoiding misunderstanding errors with the search terms; the preferences of users can be better refined and the applications can be designed to allow a better adaptation to the user; the application can communicate with other applications in order to perform in a more effective way profiting from all the information available, etc.

The aim of this book is to collect selected high-quality chapters by researchers from both academia and industry about Semantic Interoperability and related issues. The book is logically organized in three parts. The first one (Part I) has a general focus with the purpose of introducing semantic technologies and related interoperability models. Part II presents chapters in the domain of health, medicine and human behaviour modelling. Finally, Part III includes chapters focusing on systems and services.

More concretely, the book is composed of six chapters. Chapter 1 presents a deep analysis of a semantic interoperability model according to different perspectives and in the context of the most relevant technologies currently available. Chapter 2 deals with Electronic Health Records (EHR), currently considered one of the most relevant, critical and widely discussed topics in the domain of health and medicine. In Chapter 3, an analysis of the impact of semantic technologies on human behaviour modelling is presented focusing on the heterogeneity and contextual features that characterize data models and knowledge representations. Chapter 4 aims to provide a semantic perspective at industrial environment where different computation paradigms are progressively converging. A similar focus characterizes Chapter 5, deals with ontology-driven applications in smart spaces in general. Finally, Chapter 6 proposes the organization of quality-oriented data access in modern distributed environments considering a semantic interoperable ecosystem of systems and services.

1

Semantic Interoperability: Perspectives and Technology Evaluation

Raúl García-Castro[1] and Asunción Gómez-Pérez[2]

[1]*Ontology Engineering Group, Departamento de Lenguajes y Sistemas Informáticos e Ingeniería Software, Universidad Politécnica de Madrid, Spain; e-mail: rgarcia@fi.upm.es*
[2]*Ontology Engineering Group, Departamento de Inteligencia Artificial, Facultad de Informática, Universidad Politécnica de Madrid, Spain; e-mail: asun@fi.upm.es*

Abstract

Interoperability between semantic technologies is a must because they need to be in communication to interchange ontologies and use them in the distributed and open environment of the Semantic Web. However, such interoperability is not straightforward due to the high heterogeneity in such technologies. This chapter describes the problem of semantic technology interoperability from two different perspectives. First, from a theoretical perspective by presenting an overview of the different factors that affect interoperability and, second, from a practical perspective by reusing evaluation methods and applying them to six current semantic technologies in order to assess their interoperability.

Keywords: ontology engineering, ontology language, conformance, interoperability, evaluation.

S.F. Pileggi and C. Fernandez-Llatas (Eds.), Semantic Interoperability: Issues, Solutions, and Challenges, 7–24.

1.1 Introduction

Ontologies are formal specifications of shared conceptualizations that allow representing data with rich semantics [14]. In the present semantic information landscape, where ontologies and data are distributed all over the Web, semantic applications need to correctly process and interchange ontologies to manage such semantic information properly.

Due to the high heterogeneity in semantic technologies, achieving interoperability between the growing number of semantic technologies is not straightforward. Furthermore, the real interoperability capabilities of the current semantic technologies are unknown and this hinders the development of semantic systems, either when designing their interactions with other systems or when selecting the right semantic components to be used inside a system.

This converts interoperability assurance and evaluation into two key needs when developing semantic systems. These needs are currently unfulfilled mainly due to the lack of up to date information on the interoperability of existing semantic technologies.

The goal of this chapter is to describe in detail the problem of interoperability between semantic technologies from a theoretical perspective, by presenting an overview of the different factors that affect interoperability, and also from a practical one, by using existing evaluation methods and applying them to six current semantic technologies to assess their interoperability.

One of the pillars on which interoperability is based, is standards, such as the RDF(S) [2] and OWL [11] ontology language specifications defined in the W3C. Hence, the conformance of semantic technologies to these standards is a main characteristic to evaluate when gathering information about their interoperability.

Another factor that has an effect both on conformance and interoperability is the use of ontology management frameworks for processing ontologies. Using ontology management frameworks during the development of semantic systems reduces development effort and defects; however, developers should also be aware of the effect of reusing such frameworks on the interoperability of their systems.

This chapter is structured as follows. Section 1.2 provides an overview of the problem of interoperability between semantic technologies. Section 1.3 describes two approaches used to evaluate the conformance and interoperability of six different semantic technologies; the results of these evaluations are presented in Sections 1.4 and 1.5, respectively. Finally, Section 1.6 draws some conclusions from the work presented in this chapter.

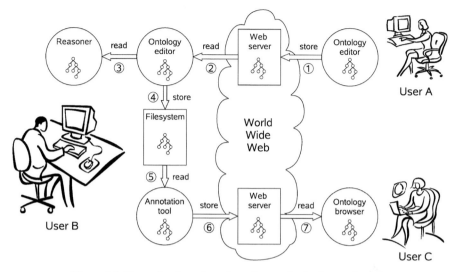

Figure 1.1 Example of ontology interchanges in the Semantic Web.

1.2 Semantic Technology Interoperability

According to the Institute of Electrical and Electronics Engineers (IEEE), interoperability is the ability of two or more systems or components to exchange information and to use this information [10]. Duval proposes a similar definition by stating that interoperability is the ability of independently developed software components to exchange information so they can be used together [3]. For us, interoperability is the ability that semantic systems have to interchange ontologies and use them.

Figure 1.1 shows an example of different ontology interchanges that could occur in the Semantic Web. In this example, a user (A) develops an ontology with his favourite ontology editor and stores the ontology in a web server. Then, a remote user (B) accesses the ontology published in the Web with his own ontology editor, makes some changes in it, and uses a reasoner for evaluating the consistency of the ontology. Afterwards, the user stores the ontology in his filesystem to later use it with an annotator to annotate his personal web page using the ontology. A third remote user (C) accesses the second user's personal web page and browses its semantic information with an ontology browser.

One of the factors that affects interoperability is heterogeneity. Sheth [13] classifies the levels of heterogeneity of any information system into information heterogeneity and system heterogeneity. In this chapter, only information heterogeneity (and, therefore, interoperability) is considered, whereas system heterogeneity, which includes heterogeneity due to differences in information systems or platforms (hardware or operating systems), is disregarded.

Furthermore, interoperability is treated in this chapter in terms of knowledge reuse and must not be confused with the interoperability problem caused by the integration of resources, the latter being related to the ontology alignment problem [4], that is, the problem of how to find relationships between entities in different ontologies.

1.2.1 Ontology Management in Semantic Systems

Ontologies enable interoperability among heterogeneous semantic systems by providing a structured, machine-processable conceptualization.

Semantic systems appear in different forms (ontology development tools, ontology repositories, ontology alignment tools, reasoners, etc.) and interoperability is a must for these technologies because they need to be in communication to interchange ontologies and use them in the distributed and open environment of the Semantic Web.

The main ontology management functionalities of semantic systems that are related to interoperability are the following:

- *Storing ontologies.* Semantic systems need to store the ontologies they use, either if these ontologies define their main data model or if they are used as information objects.
- *Operating over ontologies.* These stored ontologies need to be further processed (e.g., modified, visualised) to fulfil the system functionalities.
- *Importing ontologies.* Semantic systems need mechanisms to import ontologies coming from external systems.
- *Exporting ontologies.* Related to the previous item, semantic systems also require a means to export ontologies to other external systems.

These functionalities are usually implemented by different components in the system. Besides, when developing semantic systems it is usual to reuse components that already implement one or several of these functionalities. For example, to import and export ontologies using some ontology management framework (e.g., Jena) and even to reuse this framework for storing the ontologies.

1.2.2 Internal and External Interoperability

We can identify two types of interoperability depending on whether interoperability is required inside the semantic system limits or outside it.

- *Internal interoperability.* As mentioned above, semantic systems are frequently developed by reusing components that provide semantic capabilities and, hence, all the components inside a semantic system should interoperate correctly. Internal interoperability is achieved by means of specific software developments that guarantee interoperability and should be ensured during the development of the system.
- *External interoperability.* Semantic systems have to interact with other semantic systems to perform complex tasks, that is, semantic systems have to interoperate with external systems. External interoperability is achieved by interchanging ontologies by means of a shared resource and requires mechanisms to assess up to what extent this interoperability can be accomplished.

1.2.3 Heterogeneity in Ontology Representation

Semantic technology interoperability is highly affected by the heterogeneity of the knowledge representation formalisms of the different existing systems, since each formalism provides different knowledge representation expressiveness and different reasoning capabilities, as it occurs in knowledge-based systems [1].

This heterogeneity can be seen not only in W3C ontology language specifications where we have different ontology languages: RDF(S) [2], the OWL sublanguages (Lite, DL and Full) [11] and the OWL 2 profiles (EL, QL and RL) [12], but also in other models used to represent ontologies, such as the Unified Modeling Language[1] (UML), the Ontology Definition Metamodel[2] (ODM), or the Open Biomedical Ontologies[3] (OBO) language.

Besides, this heterogeneity does not only appear between different systems. Since different knowledge representation expressiveness and reasoning capabilities are needed for implementing distinct system functionalities, it may happen that different components inside a system have different knowledge representation formalisms.

---

[1] http://www.uml.org/
[2] http://www.omg.org/ontology/
[3] http://obofoundry.org/

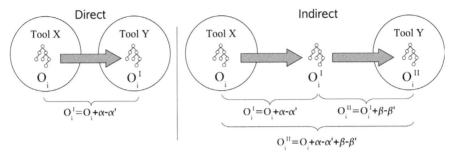

Figure 1.2 Ontology interchanges within semantic systems.

1.2.4 Aspects of Interoperability

In the beginning of this section, the IEEE definition of interoperability includes the two main aspects of interoperability, i.e., interchanging and reusing ontologies. Next, we elaborate on these aspects.

As commented above, we need to interchange ontologies, either between semantic systems or between the components of a semantic system. Two factors influence this interchange: the knowledge representation language used by the systems/components and the way of serializing ontologies during the interchange.

However, the influence of the serialization is not a big issue since, even if we can find different serializations to be used with an ontology language, it is straightforward to find a common serialization to interchange ontologies; what makes ontology interchanges problematic is the heterogeneity in knowledge representation formalisms previously mentioned.

Figure 1.2 shows the two common ways of interchanging ontologies within semantic systems: directly by storing the ontology in the destination system, or indirectly by storing the ontology in a shared resource, such as a fileserver, a web server, or an ontology repository.

The ontology interchange should pose no problems when a common representation formalism is used by all the systems involved in the interchange and there should be no differences between the original and the final ontologies (i.e., the αs and βs in the figure should be null).

However, in the real world, it is not feasible to use a single system, since each system provides different functionalities, nor is it to use a single representation formalism, since some formalisms are more expressive than others and different formalisms provide different reasoning capabilities, as mentioned above.

Most semantic systems natively manage a W3C recommended language, either RDF(S) or OWL, and sometimes both; however, some systems manage other representation formalisms. If the systems participating in an interchange (or the shared resource) have different representation formalisms, the interchange requires at least a translation from one formalism to the other. These ontology translations from one formalism to another formalism with different expressiveness cause information additions or losses in the ontology (the αs and βs in Figure 1.2), once in the case of a direct interchange and twice in the case of an indirect one.

Due to the heterogeneity between representation formalisms in the Semantic Web scenario, the interoperability problem is highly related to the ontology translation problem that occurs when common ontologies are shared and reused over multiple representation systems [9].

Finally, another aspect to take into account is that of reusing the interchanged ontology, since any transformation in the ontology could hinder its use in the destination system or prevent users from correctly using the ontology. This means that evaluating interoperability in a semantic system should involve evaluating the use of the interchanged ontologies along all the system components (and their functionalities) that make use of such ontologies.

1.3 Evaluating Semantic Technology Interoperability

The most common way for the current semantic technologies to interoperate is by means of a shared resource where the ontology is stored in a certain ontology language. In order to evaluate interoperability using an interchange language, one characteristic to consider is the conformance of the tools when dealing with ontologies defined in that language.

The next sections describe two approaches to evaluate the conformance and interoperability of semantic technologies. These approaches are motivated by the need of having an automatic and uniform way of accessing most semantic technologies and, therefore, the way chosen to automatically access the tools is through the following two operations commonly supported by most semantic tools: to import an ontology from a file, and to export an ontology to a file.

These two operations are viewed as an atomic operation. Therefore, there is not a common way of checking how good importers and exporters are; we just have the results of combining the import and export operation (the file exported by the tools). However, therefore if a problem arises in one of these steps, we cannot know whether it originated when the ontology was being imported or exported because we do not know the state of the ontology inside each tool.

Detailed descriptions of these approaches can be found in [7].

1.3.1 Evaluating Conformance

The conformance evaluation has the goal of evaluating the conformance of semantic technologies with regards to ontology representation languages, that is, to evaluate up to what extent semantic technologies adhere to the specification of ontology representation languages.

During the evaluation, a common group of tests is executed in two steps. Starting with a file containing an ontology, the execution consists in importing the file with the ontology into the origin tool and then exporting the ontology to another file.

After a test execution, we have two ontologies in the ontology representation language, namely, the original ontology and the final ontology exported by the tool. By comparing these ontologies we can know up to what extent the tool conforms to the ontology language. From the evaluation results, the following three metrics for a test execution can be defined:

- *Execution (OK/FAIL/P.E.)* informs us of the correct test execution. Its value is *OK* if the test is carried out with no execution problem; *FAIL* if the test is carried out with some execution problem; and *P.E.* (Platform Error) if the evaluation infrastructure launches an exception when executing the test.
- *Information added or lost* shows the information added to or lost from the ontology in terms of triples. We know the triples added or lost by comparing the original ontology with the final one; then we can store these triples in some human-friendly syntax (e.g., N3[4]).
- *Conformance (SAME/DIFFERENT/NO)* explains whether the ontology has been processed correctly with no addition or loss of information. From the previous basic metrics, we can define *Conformance* as a derived metric that is *SAME* if *Execution* is *OK* and *Information added* and *Information lost* are void; *DIFFERENT* if *Execution* is *OK* but *Information added* or *Information lost* are not void; and *NO* if *Execution* is *FAIL* or *P.E.*.

---

[4] http://www.w3.org/DesignIssues/Notation3.html

1.3.2 Evaluating Interoperability

The interoperability evaluation has the goal of evaluating the interoperability of semantic technologies in terms of the ability that such technologies have to interchange ontologies and use them.

In concrete terms, the evaluation takes into account the case of interoperability using an interchange language, that is, when an ontology is interchanged by storing it in a shared resource (e.g., a fileserver, a web server, or an ontology repository) and is formalised using a certain ontology language.

During the experiment, a common group of tests is executed in two sequential steps. Let us start with a file containing an ontology. The first step consists in importing the file with the ontology into the origin tool and then exporting the ontology to a file. The second step consists in importing the file with the ontology exported by the origin tool into the destination tool and then exporting the ontology to another file.

After a test execution, we have three ontologies in the ontology representation language, namely, the original ontology, the intermediate ontology exported by the first tool, and the final ontology exported by the second tool. By comparing these ontologies we can know up to what extent the tools are interoperable. For each of the two steps and for the whole interchange we have metrics similar to those presented above for evaluating conformance. Therefore, we can use the *Execution* and *Information added and lost* metrics as well as an *Interoperability* one, which explains whether the ontology has been interchanged correctly with no addition or loss of information.

1.3.3 Test Suites Used in the Evaluation

We aim to evaluate conformance and interoperability using as interchange languages RDF(S), OWL Lite and OWL DL. To this end, we will use three different test suites that contain synthetic ontologies with simple combinations of knowledge model components from these languages.

The RDF(S) and OWL Lite Import Test Suites are described in [5] and the OWL DL Import Test Suite is described in [8]. These test suites have been defined similarly in a manual way; the main difference between them is that the OWL DL test suite has been generated following a keyword-driven process that allows obtaining a more exhaustive test suite (with 561 tests compared to the 82 tests of the other test suites).

1.3.4 Running the Evaluations

The evaluations described above are part of the evaluation services provided by the SEALS Platform[5] [6], a research infrastructure that offers computational and data resources for the evaluation of semantic technologies; the mentioned test suites are also included in that platform.

Once a tool is connected to the SEALS Platform (by implementing a specific Java interface), the platform can automatically execute the conformance and interoperability evaluations. We connected six well-known tools to the platform and by means of the SEALS Platform executed the required conformance (for every tool and using every test suite) and interoperability (for every tool with all the other tools and using every test suite) evaluations.

The six tools evaluated were three ontology management frameworks (Jena, the OWL API, and Sesame) and three ontology editors (the NeOn Toolkit, Protégé OWL, and Protégé version 4).

As mentioned above, sometimes tools use ontology management frameworks for processing ontologies, which affects their conformance and interoperability. Table 1.1 shows the tools evaluated and the ontology management frameworks (i.e., APIs) that they use, including in both cases their version numbers.

1.4 Conformance Results

This section presents the conformance results for the six tools evaluated. Tables 1.2, 1.3 and 1.4 present the tool conformance results for RDF(S), OWL Lite and OWL DL, respectively.[6] The tables show the number of tests in each category in which the results of a test can be classified, depending on whether the original and the resultant ontologies are the same (*SAME*), are different (*DIFF*), or the tool execution fails (*FAIL*).

As can be observed in these tables, Jena and Sesame present no problems when processing the ontologies included in the test suites for the different languages. Therefore, no further comments will be made on these tools.

Besides, as shown in Table 1.1, the NeOn Toolkit and Protégé 4 use the OWL API for ontology management.

The version of Protégé 4 evaluated uses a version of the OWL API that is almost contemporary to the one we evaluated. Hence, after analysing the

---

[5] http://www.seals-project.eu/seals-platform

[6] The tool names have been abbreviated in the tables: JE = Jena, NT = NeOn Toolkit, OA = OWL API, P4 = Protégé 4, PO = Protégé OWL, and SE = Sesame.

Table 1.1 List of tools evaluated.

Ontology management frameworks			
Tool	**Version**		
Jena	2.6.3		
OWL API	3.1.0 1592		
Sesame	2.3.1		
Ontology editors			
Tool	**Version**	**API**	**API version**
NeOn Toolkit	2.3.2	OWL API	3.0.0 1310
Protégé 4	4.1 beta 209	OWL API	3.1.0 1602
Protégé OWL	3.4.4 build 579	Protégé OWL API	3.4.4 build 579

Table 1.2 RDF(S) Conformance results.

Category	JE	NT	OA	P4	PO[a]	SE
SAME	82	0	0	0	68	82
DIFF	0	82	82	82	14	0
FAIL	0	0	0	0	0	0
TOTAL	82	82	82	82	82	82

[a]Not counting additions of *owl:Ontology*.

Table 1.3 OWL Lite Conformance results.

Category	JE	NT[a]	OA[a]	P4[a]	PO	SE
SAME	82	78	80	80	73	82
DIFF	0	2	2	2	9	0
FAIL	0	2	0	0	0	0
TOTAL	82	82	82	82	82	82

[a]Not counting additions of *owl:NamedIndividual*.

Table 1.4 OWL DL Conformance results.

Category	JE	NT[a]	OA[a]	P4[a]	PO	SE
SAME	561	549	549	549	429	561
DIFF	0	8	11	11	132	0
FAIL	0	4	1	1	0	0
TOTAL	561	561	561	561	561	561

[a]Not counting additions of *owl:NamedIndividual*.

results of Protégé 4 we reached the same conclusions that those obtained for the OWL API and the comments made for the OWL API are also valid for Protégé 4.

However, the version of the NeOn Toolkit evaluated uses a version of the OWL API that differs in some months to the one we evaluated. In general, from the results of the NeOn Toolkit we reached the same conclusions that those obtained from the OWL API. In the next sections we will only comment on those cases where the behaviour of the NeOn Toolkit and the OWL API differ.

1.4.1 RDF(S) Conformance

When the *OWL API* processes RDF(S) ontologies, it always produces different ontologies because it converts the ontologies into OWL 2. The changes performed over the ontologies are the following:

- Classes are transformed into OWL classes.
- Individuals are transformed into OWL 2 named individuals.
- Properties are transformed according to their use in the ontology. If a property either relates two classes or two individuals or has as domain a class and as range another class (even if the range class is *rdfs:Literal* or an XML Schema Datatype), it is transformed into an OWL object property. If a property either relates one individual with a literal value or does not have domain and has a range of *rdfs:Literal* or an XML Schema datatype, it is transformed into an OWL datatype property. If conditions from these two groups appear in an ontology (e.g., a property relates one individual with a literal value and has as domain a class and as range another class), the property is created both as an object property and a datatype property.
- Classes and properties that are described no further in the ontology (i.e., the only statement made about a resource is that it is a class or a property) are lost.
- Undefined resources that either appear as domain or range of a property or that have an instance are defined as OWL classes.
- Classes related by a property or classes related to a literal value using a property are transformed into OWL 2 named individuals.
- A particular instance of the previous case is that of metaclasses (i.e., when the property that relates the two classes is the *rdf:type* property). In this case, classes without instances and that are instance of another class are transformed into OWL 2 named individuals. Besides, classes that have instances and that are instance of another class are defined both as OWL classes and as OWL 2 named individuals.

When *Protégé OWL* processes an RDF(S) ontology, the ontology is always created as an OWL ontology with a randomly generated name.[7] Re-

---

[7] E.g., http://www.owl-ontologies.com/Ontology1286286598.owl

gardless of this, the cases when different ontologies are produced occur when the ontology contains

- A property with an undefined resource as range. The undefined resource is created as a class.
- A literal value. The literal value is created with a datatype of *xsd:string* and, therefore, it is a different literal. According to the RDF specification, one requirement for literals to be equal is that either both or neither have datatype URIs.[8]

1.4.2 OWL Lite Conformance

When the *OWL API* processes OWL Lite ontologies, it converts the ontologies into OWL 2. Since OWL 2 covers the OWL Lite specification, most of the times the OWL API produces the same ontologies. However, one effect of this conversion is that individuals are converted into OWL 2 named individuals.

The cases in which the ontologies are different occur when the ontology contains a named individual related through an object property to an anonymous individual, and this anonymous individual is related through a datatype property to a literal value. In this case, the named individual is related through the object property to an anonymous resource, another anonymous resource is related through a datatype property to a literal value, and the anonymous individual is not related to anything.

After analysing the results of the *NeOn Toolkit* we obtained the same conclusions that were previously presented for the OWL API with one exception. When the ontology contains an anonymous individual related to a named individual through an object property, the execution of the NeOn Toolkit fails.

When *Protégé OWL* processes an OWL Lite ontology, most of the times it produces the same ontology. The only exception is when the ontology contains a literal value. The literal value is created with a datatype of *xsd:string* and, therefore, it is a different literal. As mentioned above, one requirement for literals to be equal is that either both or neither have datatype URIs.

1.4.3 OWL DL Conformance

When the *OWL API* processes OWL DL ontologies, it converts the ontologies into OWL 2. Since OWL 2 covers the OWL DL specification, most of the

---

[8] http://www.w3.org/TR/2004/REC-rdf-concepts-20040210/
#dfn-typed-literal

times the OWL API produces the same ontologies. However, one effect of this conversion is that individuals are converted into OWL 2 named individuals.

The cases when the ontologies are different occur when the ontology contains

- An anonymous individual related through an object property to some resource or through a datatype property to a literal value. In this case, an anonymous resource is related through the property to the resource or literal, and the anonymous individual is not related to anything.
- A datatype property that has as range an enumerated datatype (i.e., an enumeration of literals). In this case, an *owl:Datatype* class is created as well as an anonymous individual of type *owl:Datatype*. However, the *owl:Datatype* class does not exist in the RDF(S), OWL or OWL 2 specifications; only the *rdfs:Datatype* class exists.

There is another case in which the test execution fails; when the ontology imports another ontology, the OWL API does not produce any ontology. This happens because, as the OWL API cannot find the ontology in the *owl:imports* property, it does not import anything. However, the tool should not rely on having full access to an ontology for just importing such ontology.

After analysing the results of the *NeOn Toolkit* we obtain the same conclusions as those previously presented for the OWL API with one exception; when the ontology contains an anonymous individual related to another anonymous individual through an object property, the execution of the NeOn Toolkit fails.

When *Protégé OWL* processes OWL DL ontologies, it usually produces the same ontology. The cases in which the ontologies are different occur when the ontology contains

- A literal value. The literal value is created with a datatype of *xsd:string* and, therefore, it is a different literal. As mentioned above, one requirement for literals to be equal is that either both or neither have datatype URIs.
- Class descriptions that are the subject or the object of an *rdfs:subClassOf* property. In these cases, the class description is defined as equivalent to a new class named "Axiom0"; this new class is the subject or the object of the *rdfs:subClassOf* property.

Table 1.5 RDF(S) Interoperability results.

	JE	SE	PO	NT	OA	P4
JE	100	100	83	0	0	0
SE	100	100	83	0	0	0
PO	83	83	83	0	0	0
NT	0	0	0	0	0	0
OA	0	0	0	0	0	0
P4	0	0	0	0	0	0

Table 1.6 OWL Lite Interoperability results.

	JE	SE	OA	P4	NT	PO
JE	100	100	98	98	95	89
SE	100	100	98	98	95	89
OA	98	98	98	98	95	89
P4	98	98	98	98	95	89
NT	95	95	95	95	95	87
PO	89	89	89	89	87	89

Table 1.7 OWL DL Interoperability results.

	JE	SE	OA	P4	NT	PO
JE	100	100	98	98	98	76
SE	100	100	98	98	98	76
OA	98	98	98	98	98	75
P4	98	98	98	98	98	75
NT	98	98	98	98	98	75
PO	76	76	75	75	75	76

1.5 Interoperability Results

This section presents the interoperability results of the six tools evaluated. Tables 1.5, 1.6 and 1.7 present the tool interoperability results for RDF(S), OWL Lite and OWL DL, respectively.[9]

The tables show the percentage of tests in which the original and the resultant ontologies involved in an interchange are the same. For each cell, the row indicates the tool origin of the interchange, whereas the column indicates the tool destination of the interchange.

---

[9] As in the conformance tables, in these tables we have not counted additions of *owl:NamedIndividual* and *owl:Ontology*.

The conclusions of the behaviour of the tools that can be obtained from the interoperability results are the same as those already presented when analysing their conformance. The only new fact obtained while analysing the interoperability results stems from the interchanges of OWL DL ontologies from the OWL API (or from those tools that use the OWL API, i.e., Protégé 4 and the NeOn Toolkit) to Protégé OWL. In these interchanges, when the ontology contains an anonymous individual related through a datatype property to a literal value, Protégé OWL has execution problems.

In summary, in terms of interoperability

- Regardless of the ontology language used in the interchange, Jena and Sesame have no interoperability problems whereas the rest of the tools suffer some issues that prevent their full interoperability.
- Tools based in the OWL API convert RDF(S) ontologies into OWL 2 and, hence, they cannot interoperate using RDF(S) as the interchange language.

1.6 Conclusions

This chapter has presented an overview of the problem of interoperability between semantic technologies, first from a theoretical perspective and, second, by evaluating some tools and describing some of the problems the tools encounter when processing and exchanging ontologies.

In the results we can observe that all the tools that manage ontologies at the RDF level (Jena and Sesame) have no problems in processing ontologies regardless of the ontology language. Since the rest of the tools evaluated are based in OWL or in OWL 2, their conformance and interoperability is clearly better when dealing with OWL ontologies.

From the results presented in the previous section we can note that conformance and interoperability are highly influenced by development decisions. For example, the decision of the OWL API developers (propagated to all the tools that use it for ontology management) of converting all the ontologies into OWL 2 makes the RDF(S) conformance and interoperability of these tools quite low.

Since the OWL Lite language is a subset of the OWL DL, there is a dependency between the results obtained using the test suites for OWL Lite and OWL DL. In the results we can also observe that, since the OWL DL test suite is more exhaustive than the OWL Lite test suite, the OWL DL evaluation unveiled more problems in the tools than the OWL Lite evaluation. These problems included not only issues related to the OWL DL language, but also some related to OWL Lite ontologies involved in the OWL DL test suite.

The results also show the dependency between the results of a tool and those of the ontology management framework that the tool uses; using a framework does not isolate a tool from having conformance or interoperability problems. Besides inheriting existing problems in the framework (if any), a tool may have more problems if it requires further ontology processing (e.g., its representation formalism is different from that of the framework or is an extension of the framework formalism) or if it affects the correct working of the framework.

However, using ontology management frameworks may help increase the conformance and interoperability of the tools, since developers do not have to deal with the problems of low-level ontology management. Nevertheless, as observed in the results, this also requires to be aware of the defects contained in these frameworks and to regularly update the tools and thus use their latest versions.

Acknowledgements

This work has been supported by the SEALS European project (FP7-238975). Thanks to Rosario Plaza for reviewing the grammar.

References

[1] R. Brachmann and H. Levesque. A fundamental tradeoff in knowledge representation and reasoning. In *Readings in Knowledge Representation*, pages 31–40. Morgan Kaufmann, San Mateo, 1985.

[2] D. Brickley and R.V. Guha (Eds.). RDF Vocabulary Description Language 1.0: RDF Schema. W3C Recommendation, 10 February 2004.

[3] E. Duval. Learning technology standardization: Making sense of it all. *International Journal on Computer Science and Information Systems*, 1(1):33–43, 2004.

[4] J. Euzenat and P. Shvaiko. *Ontology Matching*. Springer-Verlag, 2007.

[5] R. García-Castro. *Benchmarking Semantic Web technology*, volume 3 of *Studies on the Semantic Web*. AKA Verlag–IOS Press, January 2010.

[6] R. García-Castro, M. Esteban-Gutiérrez, and A. Gómez-Pérez. Towards an infrastructure for the evaluation of semantic technologies. In *Proceedings of the eChallenges 2010 Conference*, Warsaw, Poland, October 27–29, 2010.

[7] R. García-Castro, S. Grimm, M. Schneider, M. Kerrigan, and G. Stoilos. D10.1. Evaluation design and collection of test data for ontology engineering tools. Technical report, SEALS Project, November 2009.

[8] R. García-Castro, I. Toma, A. Marte, M. Schneider, J. Bock, and S. Grimm. D10.2. Services for the automatic evaluation of ontology engineering tools v1. Technical report, SEALS Project, July 2010.

[9] T.R. Gruber. A translation approach to portable ontology specifications. *Knowledge Acquisition*, 5(2):199–220, 1993.

[10] IEEE-STD-610. *ANSI/IEEE Std 610.12-1990. IEEE Standard Glossary of Software Engineering Terminology*. IEEE, February 1991.

[11] D.L. McGuiness and F. van Harmelen. OWL Web Ontology Language Overview. W3C Recommendation, 10 February 2004.

[12] B. Motik, B. Cuenca-Grau, I. Horrocks, Z. Wu, A. Fokoue, and C. Lutz. OWL 2 Web Ontology Language Profiles. W3C Recommendation, 27 October 2009.

[13] A. Sheth. Changing focus on interoperability in information systems: From system, syntax, structure to semantics. In *Interoperating Geographic Information Systems*, pages 5–30. Kluwer, 1998.

[14] R. Studer, V.R. Benjamins, and D. Fensel. Knowledge engineering: Principles and methods. *IEEE Transactions on Data and Knowledge Engineering*, 25(1–2):161–197, 1998.

PART II

Health, Medicine and Human Behavior Modeling

2

Electronical Health Records (EHR) Systems Interoperability and Accessibility: ISO 13606 and Ontologies

Belen Prados Suárez[1], Carlos Molina[2],
Miguel Prados[3] and Carmen Peña[3]

[1] Software Engineering Department, University of Granada, Spain;
e-mail: belenps@ugr.es;
[2] Computer Science Department, University of Jaen, Spain;
e-mail: carlosmo@ujaen.es;
[3] Computing Department, University Hospital San Cecilio, Spain;
e-mail: {prados, carmenpy}@decsai.ugr.es

Abstract

In this chapter we present a proposal to conceptualize the EHR, based on the semantic description of the information, according to the documentary structures and the clinical aspects of the EHR contents. Our aim here is to perform a formalization with a double purpose: on one hand to enable the interoperability based on ISO 13606; on the other hand, to improve the accessibility to the EHR, according to clinical or assistance contexts, providing the clinical data retrieval system with flexibility and operativity. To this purpose we propose the use of an ontology to represent this conceptualization, and include properties and relations between the components of the EHR. To do it, we integrate the ISO 13606 as an ontology in the global EHR system structure so the interoperability is inside the system and not as an interface. The main result is an EHR structure that improve the access using the semantic of the clinical contexts and allow the communication of information with other systems.

S.F. Pileggi and C. Fernandez-Llatas (Eds.), Semantic Interoperability: Issues, Solutions, and Challenges, 27–47.

Keywords: electronic health records, interoperability, ISO 13606, accessibility, ontology.

2.1 Introduction

Every day the Electronic Health Record (EHR) becomes more of an extended reality in the majority of the hospitals, with different degrees of development. It has opened the way to new uses of the EHR, optimized and with more benefits for the medical activity. However, new problems and perspectives have also arisen, related to the management of the clinical information [26]. As the use of the EHR spreads over the different medical specialities, it must integrate more documents and information items, from different sources and types. It is unavoidable to think about the risk that the EHR runs of becoming as unmanageable as the old health records on paper: with such a quantity of information and documents, the access to concrete data items required in relatively simple situations can be really difficult. Another main problem is interoperability, where the aim is to communicate and make possible the understanding between different models of EHR from different hospitals and providers. The ISO 13606 [17] regulation establishes the basis and general framework of the semantic interoperability model [2, 11, 19, 29], to allow the univocal interpretation of the information transmitted within the context in which it was generated. The ISO 13606 regulation proposes a dual model where the first model is the reference model and the second one is the archetypes model. Both of them will be explained in this chapter. In addition to the above mentioned problems, there are also several important issues that must also be addressed and solved, like:

- *The Mobility*: The use of mobile devices (tablet PC, PDAs, etc.) requires agile and summarized navigation models on the EHR.
- *The Contextualization*: The contextual use of the information would provide the doctor with information pertinent to the assistance act where he/she is involved.
- *Access Focusing*: The idea is to allow the navigation through the EHR directing the search according to a semantic purpose.
- *The personal access of the owner*: Patients increasingly demand access to their clinical data since, as owner of them, they have the right to access them. However, at this moment neither are the systems ready for this purpose nor do the citizens have the technical knowledge to use them.

The literature reviewed shows a great concern about these problems of the EHR systems, specially interoperability, and in particular about the use of ontologies as a means to represent models capable of understanding and communicating to each other [14]. However, though in the literature most of the proposals are mainly focused on the interoperability and its tools, they do not propose explicitly applications towards the accessibility, use and management of the EHR at a local level.

In this chapter we offer a different point of view, more focused on this latter line, offering different alternatives of use of the EHR for the doctors, making the accessibility to the information needed more efficient. This is quite important since the volume of documents and information contained in the EHRs is so extensive that usually one of the main causes of complains from the users is related to the difficulties to navigate through them.

This leads us to think that the EHR can be considered as a universe of knowledge that can be conceptualized in such a way that each individual information item can be defined in a semantic family, according to its properties and relations with other information items [15]. In addition, this chapter provides a novel point of view, applying the ontology based representation models to make the use and navigation easier.

It is specially interesting, since it opens the possibility of performing semantic retrieval of information, through the construction of agents that interpret the queries and allow access to the pertinent data items.

Our proposal then is an approach to the conceptualization of the universe of clinical data contents of the EHR, through the use of an ontology, based on our own EHR model. The proposal will be included inside a more complex system (the complete Hospital Information System) which has been modelled using ontologies to improve user accessibility.

Starting from this conceptual formalization and on the basis of the generated knowledge we can construct "agent procedures". These procedures guide the user to those information items that, due to their semantic value, can be more useful according to the access model.

2.2 Related Work

Nowadays the problem of interoperability is an important open issue. In the last decades, much work has been done to solve how one hospital can access the information that another has about a patient. The main problems are caused by the different information systems (most of the Hospital Information Systems have been developed ad-hoc for the concrete needs of each hospital) and the different regulations (e.g. regional regulations [10]) so there is a huge variety of architecture and proposals. These systems stored not only the information about patients but also information about the operational aspect of the hospitals. The part that manages patients' information is the equivalent of the traditional Health Records, so these system are called Electronical Health Records Systems (or EHR [5, 7]).

There have been proposals to unify the terms and the meaning in these contexts [1, 6, 13] and for representing some types of information (e.g. the standard DICOM for medical images [16]). But a more deep consensus is needed to allow the systems to communicate with each other. The problems of interoperability have been studied and the archetypes are one of the most promising proposals [3, 8, 10]. Although there is no standard, there are some proposals for the organization of these systems using the archetype-concepts as OpenEHR [22]. In the last years a new standard has appeared: ISO 13606 [17]. This standard is also based on archetypes but is less restrictive than OpenEHR so only proposes an interface structure between systems and not a complete architecture as OpenEHR. In that respect, the use of ontologies may help in the development of systems and the sematinc interoperability between systems (e.g. [18, 21, 27, 29]). Most of the proposals are based on the use of ontologies to represent the system information or as a layer to allow the communications between systems. In all the cases these are ad-hoc proposals that are not compatible with the standards.

In this chapter we introduce a point of view considering the standard ISO 13606 as a "simplified" ontology (it is an object hierarchy only) and integrating this ontology into the complete system modelled as an ontology.

2.3 Background

Before presenting our proposal we shall make some notes about the framework from which we have started our research, specially the information system and the legal framework. To focus the chapter we need to establish the base system for the research and the developments.

2.3.1 Electronic Health Record (EHR)

In this section, we present a brief definition about the Electronical Health Records and some information about the database that stores them.

We have based our study on the Electronic Health Record System of the University Hospital San Cecilio in Granada, which stores around 1,000,000

EHR, containing more than 50 million documents. In the future, it is expected to vastly increase in size, due to the inclusion of new types of documents from two sources: old documents that still have not been digitalized (scanned images, MRI, etc.) and new documents generated by the recently acquired devices and equipment like PACS (Picture Archiving and Communications System: systems to obtain images from medial equipment like radiographies, and to visualize and to store them in the EHR) and other devices that will become available in the near future.

2.3.1.1 Logical Organization of EHR

To better understand our proposal, it is necessary to know the logical organization of the data in the EHR.

The information stored in the EHR is grouped into logical units called *documents*, where each one represents a set of data with a uniform informative clinical guidance. Inside the documents, there are data that can also be grouped into small logical units, which we call *data groups*, when they are related under a clinical point of view (Figure 2.1). Each document is characterized by a set of properties [26, 28] like the type (exploration, anamnesis, epicrisis, checkup, nursing control, intervention, external, etc.), the specialty (medical specialty as surgery, cardiology, and so on, nursing, administrative, etc.), or the pathological or clinical process (documents about pregnancy, cataracts, diabetes, etc.). Each data group inherits these general properties from the document it is in. In addition, each data group has its own specific properties, as the relevance level (for the concrete patient and episode), the confidentiality level, etc.

As an example of the existing documents, we have a blood analysis or a preanaesthetic study; and examples of data items are: erythrocyte, hemoglobin, corpuscular volume, amylase, GGT, HDL-cholesterol, LDL-cholesterol, or VLDL-cholesterol, in the first case; and hypertension, cardiopathy, electrocardiogram, radiologic study, or echography, in the second case. These data items can be grouped into the data groups general "biochemistry" and "lipid information", for the first type of document; and "risk factors" or "additional tests", for the second type. Here we would like to remark that the identification information of the patient and EHR is a "special" data group, since it is common to all documents, so we do not take it into account for the processes that we will explain later. For all the data groups the system uses the ICD-10 diagnosis enconding [24] when applicable.

In our system, documents are organized according to *assistance episodes* (e.g., admissions, outpatient consultation, emergency assistance, day

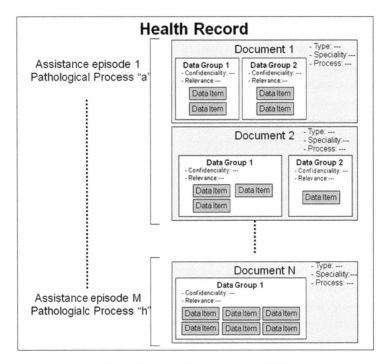

Figure 2.1 Logical organization of EHR.

hospital) in a chronological or medical ordering, depending on the assistance processes. Documents are classified according to the types, considering 1,500 different document types in the system: intervention sheet, progress sheet, nursing sheet, pregnancy process, diabetes protocol, radiological report, and so on.

This logical organization allows the processing and analysis of the information, both at *document* level and at individual *data item* level or *data group* level. Here, we consider the data groups as the minimum unit of information, since a single data item can be managed as a data group with just one element.

2.3.1.2 EHR Database

The structure of the system is shown in Figure 2.2. The users access the system using medical workstations. These are normal PCs, light PCs (or net PCs) or the most recently incorporated terminals like Tablet PCs and PDAs. The user then logs onto the system and gains access to a Citrix [12] farm of

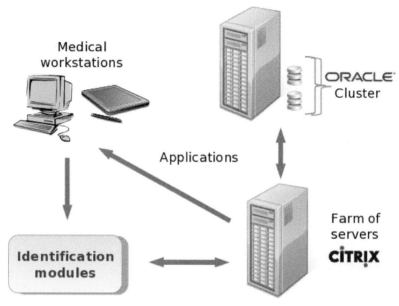

Figure 2.2 Structure of the system.

servers where the applications are executed. All the data are stored in a data base cluster using Oracle DBMS [23]. The system integrates software from different providers (adapted to the particular system structure and needs) as well as software developed in-house.

This system, as legally required, stores each access to the EHR, indicating the accessed data and, in case of modification, the modified data; the staff member who accesses such data; and the assistance situation (called controlled assistance situation) in which the access occurs. It is called *Retrospective Access Base (RAB)*. The number of records stored in the RAB is in the order of hundreds of millions. Due to the size, it is stored in the database of the system (Oracle).

2.3.2 Ontology-based Hospital Information System

Nowadays one characteristic of Hospital Information Systems (HIS) is the huge amount of data they manage, generate and store, as well as their wide variety and typology. Among these data reports or outputs it can be found which are the direct result of concrete queries, as well as results, that are more elaborated data obtained from statistical processes or a basic exploitation of the information in the DB, using the tools offered by the applications of the concrete area or department [26].

Examples of reports are the list of patients waiting for a given surgery or the list of medicines prescribed by each medical specialty; while the average duration of the hospital stay or the efficiency indicators in each medical specialty are examples of results. The set of all these documents and data is what we call Universe of Result Reports (URR [29]).

A serious inconvenience is that the access to these data must be done by a complicated browse though the menus of different applications, which gives rise to severe problems like:

- There is so much information available that the user cannot find what he/she is looking for.
- There are a number of reports and outputs that are wrongly classified, and these are sometimes useless.
- There is no proper organization of the information, making the retrieval quite difficult.
- There is also a great amount of redundant results.
- The user does not know the structure of the results, and hence what can he/she obtain from them and which application does what.
- The user obtains something that is not what he/she really needs, or even cannot find it.
- The models used to exploit the information change over time, as well as the criteria, purposes and requirements.

To solve these problems it is necessary to define logical structures to organize the URR, designing it under semantic criteria. In this way new navigation systems can be designed to allow the user to focus or widen the information queries [18].

We have developed an ontology to establish a conceptualization of the URR. Our proposal is to integrate in this ontology the needed classes to enable the interoperability of the EHR (a subsystem in the HIS). We first present the ontology for the internal access and later we will extend it with the classes for the interoperabiltiy.

The actually implemented HIS covers the economical-administrative and logistical-assistance fields, managing the operational systems of the hospital organization, and hence its activity.

Functionally it is structured in several levels, as indicated in Figure 2.3.

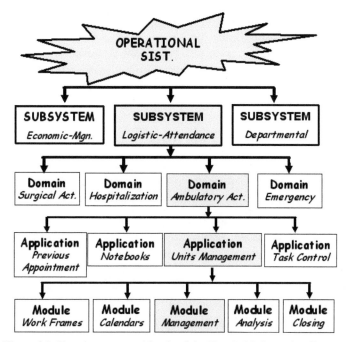

Figure 2.3 Function structural levels of the Hospital Information System.

2.3.3 Formal Analysis of the URR

Formally a document from the URR has the following characteristics:

- *Type:* This property can have one of the next values: "query relation report", "operative recount" or "statistical report".
- *Proprietary Application:* This characteristic references the functional module from the HIS logical design that manages the information and procedures related to the concrete target document. This characteristic follows the logical levels structure of the HIS, shown in bold letters in Figure 2.3 (System – Subsystem – Functional Scope – Application – Module – Procedure). Its values are, for example, "absenteeism" Module, from the "Human Resources Management" Application, as well as those in capital letters in Figure 2.3.
- *Computing procedure that generates (executes) the document:* This attribute is referred to the physical unit from the physical design that generates the target document.
- *Domain of the showed attributes:* This is the set of data from the Operational System that are showed in the document.

- *Generated Data:* References the total results or statistical values computed on the execution process.
- *Query descriptors:* Are those that allow the user to filter the query for a particular document. Usually they are very limited sets like date, functional unit, diagnosis, etc., and can be structured according to their meaning.

2.3.4 Semantic Analysis of the URR

The semantic analysis of the URR offers a perspective about its meaning from a given point of view of interest, like the "business management" perspective from the point of view of the "administrative management".

This task acquires maximum interest if we consider that we have a logical structure of the knowledge, shown in Figure 2.4, based on several levels (contexts, scenarios, images, views), which makes the semantic pertinency of a given document to a concrete image or view possible [25].

As an example, a report about the occupation of the operating rooms is related (pertinent) to the "efficiency of operating rooms use" image, which belongs to the "surgical activity" scenario from the "medical care" context.

In addition to the pertinency, there may be other characteristics with semantic values like:

- Orientation: determines the type of user to which the document is directed: Medical directors, Management, Central Services, etc.
- Confidentiality level: with three possible values: low, medium and high.

2.3.5 Structure of the Ontology

The aim of the ontology design is to elaborate a proposal according to the existing HIS, but open to the incorporation of new elements to the URR, that could give rise to new "main classes" (defined later) or instances. The basic criteria that must be verified are hierarchization (ordering), modularization (concept isolation), abstraction, clarity and coherency. Considering this, we have defined the following elements for the design:

- Domain: Set of reports and results (URR) that can be obtained from the Hospital Information System. Hence, the concepts or domain elements will be the results, corresponding to reports/outputs/lists issued from the variety of information sources, on user demand.
- Classes: Each class corresponds to one of the formal or semantic characteristics defined on the URR. Here we define the concept of "main

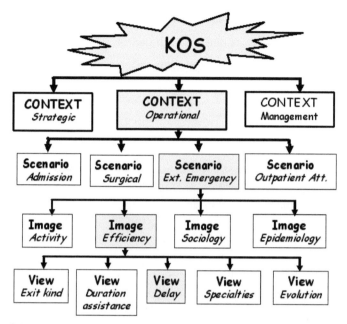

Figure 2.4 Semantic categorization scheme. Example: operational context scenarios.

class" to mean that the instances of that type of class are elements from the URR, and we establish as main class the semantic categorization.

This way, as an example we say that an "external emergency assistance delay report" is an instance of the class "emergency efficiency", which is a subclass of "emergency assistance activity", instance at the same time of "assistance activity" that is a subclass of the "operational context".

- Properties: Assigned to each element or class of the URR domain, represent the set of characteristics of a given element or set of elements. There are two types of properties: complex properties, defined as relations between ontology classes; and explicit properties, defined exclusively for a given class or instance, without generating a new class.
- Instances: the elements in the URR are assigned to instances of the semantic categorization. They will be represented by the naming of the physical unit of the HIS. There will also be instances corresponding to the individual terminal elements of the class hierarchies, and will be defined in the ontology as "direct classes".

2.3.6 Identification of Ontology Classes

The main conceptual classes defined in the ontology are:

- *Semantic Categorization:* Represents the cognitive focus approached by a URR document, and obeys to user criteria formalized following the levels schema in Figure 2.4. Its instances are the URR documents expressed as a code according to the physical unit or executable computing procedure.
- *Logical Model Functions:* This class represents the organization activities, computerized and structured according to hierarchies of functional requirements. An approach to this class hierarchization is shown in Figure 2.3. As an example, the control of the "temporary working inability" is a member of "absenteeism control", member at the same time of "human resources management", member again of "administrative management", that belongs to the economic-administrative subsystem.
- *Receiver:* final destination of the document. We have planned three types: "local", "external" and "publication". The class "local" generates a hierarchic line following the organic structure of the hospital government. The "external" class represents different institutions that are users of the hospital information, while "publication" refers to different ways to announce the information (web pages, notice board, and so on).
- *Chronology:* establishes the "periodical" (and period) or "on demand" character.
- *Confidentiality Level:* unavoidably associated with every hospital document.
- *Query Strategy:* represents the dimensions through which the user can select or filter the content of a document. It is useful to inform the user about the possibilities to focus the query.
- *Leaning to decision:* indicates the interest of the document under a strategic point of view, like periodic control, evolution or research.
- *Semantic Valoration of the Contents:* refers to the set of resulting information in a URR document. Concretely, it refers to the physical data model, and the structure and contents of the data tables themselves. Hierarchies in this class are identified according to three levels: "application", "table type" (historical, movements, master, and base) and "data type" (own attribute, situation attribute, movement attribute).

2.3.7 Properties Definition

The aim of the properties is to enrich the conceptual structure of the ontology, to properly define URR documents according to their formal and semantic characteristics. We have two types of properties:

- *Complex Properties:* are an integral part of the ontology, and are defined as relations between classes. As an example, a report of the "attended emergencies", which is a member of "emergency frequentiation", has the properties of being confidential, of local use for Medical management, can be selected chronologically, functionally or by its characters, and show the data in the table "emergency movements".
- *Explicit Properties:* represent qualities of one or several classes, and are useful only when they are assigned to a few classes since, otherwise they would finally become a new class. This type of properties is reserved for very specific characteristics of a class or instance, like a document which totals some concrete values according to a specific calculation. These properties can be easily added and removed to enable or disallow some given characteristic.

2.4 ISO 13606

As mentioned above, the ISO 13606 [17] regulation proposes a dual model where the first model is the reference model and the second one is the archetypes model. The proposal tries to define a general structure for EHR system interfaces. So, the proposal is only for interoperability but not for the intenal structure of the system. The ISO 13606 proposes a hierachical object structure to classify and store the medical concepts (e.g. diseases, reports, etc.) and the use of archetypes for each of these concepts. It is based on other proposals such as Open-EHR and the requirements by companies related to health. The ISO proposes to use a message using HL 7 version 3 to communicate between the systems. The agents implicated in these messages are not only EHR systems but also other middleware services such as security components, workflow systems, alerting and decision support services and other medical knowledge agents.

2.4.1 Reference Model

The reference model is proposed to structure the data. It established a basic structure using an object-oriented paradigm. It defines the main classes with the characteristics to store on each one. The classes are structured in a hierarchical manner from a set of documents (folder) to each value on a single analysis. It is based on a class called "structure" that gives rise to the following hierarchy of members:

- Folder: represents the divisions at the highest level inside the extracts of the clinical history.
- Composition: is the set of annotations related to a unique given clinical session or document.
- Sections: are groupings in a clinical session.
- Entry: each one represents a clinical observation or a set of them.
- Cluster: is used when the representation of a unique observation or action requires a complex data structure, like a list, a table or a temporal series.
- Element: contains a unique value that must be an instance of some of the types defined by it.

2.4.2 Archetypes

The second model describes the Archetypes [3, 4, 8, 10, 30] as a way to define the clinical concepts managed by the system. The archetypes are definitions of sets of clinical information items, that have a concrete clinical meaning; and they are created using the components defined in the ISO 13606.

Examples of these archetypes may be:

- Pathological processes: cataract.
- Protocols: pregnancy.
- Documents: blood analysis, radiography and prescription.
- Archetype can also be a group of items (biochemistry or lipid information) or a single item (HDL-cholesterol, LDL-cholesterol or VLDL-cholesterol) in a document.

So an archetype is any information item or group of items related to a clinical point of view.

However, this regulation just sets the basis and general description on which everything is opened and must be concretized, which is what we do in this chapter.

2.5 Ontology

To realize the general models described in the previous section, we propose and describe in this section a new ontology that integrates the HIS structure and the ISO 163003 proposal. This ontology will alow both the use of the system in the hospital and the interoperability based on the standard.

2.5.1 General Description

The EHR structure's design itself implies the existence of a categorization according to the semantic classes of the documentary organization, and also to the assistance part. As an example, the documents are classified by their types, and the data are organized regarding their clinical orientation inside the document. In addition, these data items are organized according to assistance acts and medical specialities. Based on this, we make the formalization through the ontology. To choose the components of the ontology we have used two criteria. On the one hand, the documentary criteria, that gives rise to the classes that structure and define the set of documents included in the EHR. On the other hand, we approach clinical criteria determinating the categories related to the clinical processes and concrete pathologies, and even the assistance context on which the information is used. Nevertheless, the semantic universe must be manageable and easily formalizable, so we have avoided to define categories that are not clearly useful to reach our final targets, previously indicated.

2.5.2 Semantic Categories in the Ontology

We have analyzed the semantic categories to define the ontology. The main purpose was twofold: on the one hand, to categorize every kind of data that could be found in the EHR, on the other hand, to respect the information structures defined in our EHR model. With it the classes defined in the ontology are:

- Structure Model (EHR-EXTRACT): This class corresponds to the structure defined in the ISO 13606, and has the members Folder, Composition, Section, Entry, Cluster and Element.
- Document: A document can be considered as any grouping of data with a common purpose, nested regarding a clinical action or observation. The documents are hierarchized depending on whether they are "general", "process", "medical speciality", "nursery", "surgical" or "logistical". Hence, this class can be considered as the fundamental logical group-

ing of the organization of the information in the EHR. With this class
the EHR can be organized according to assistance acts (admissions,
consultations, emergencies, etc.) or to pathological processes, always
grouping documents. Each document may contain different sections of
contents, and each section has its own entries, clusters and elements as
concrete data in the document.

- Assistance Process: These processes define the clinical pathology environments, previously set, on which sequences of clinical actions are
  pre-established. As an example, we have the "diabetes process" and the
  "cataract process". Here we have focused on the pathologies with well
  defined processes, since not all the pathologies have them. The members of this class represent the different pre-established actions for each
  process.
- Data Type: They can be considered as texts, encoded data, mag-
  nitudes that include rations, intervals, lengths, durations, graphs, images,
  signals, dates and so on.
- Observation Type: The aim of this class is to qualify the data item
  according to its source: if it is a subjective observation, an objective
  result of an analysis, a protocolyzed observation, a related fact or a
  chronological action, among others.
- Assistance Procedure: It contains the references to the diagnosis methods, explorations, sources of knowledge, technological support, and
  any other source of data. As an example, we have electromedical
  explorations (electrocardiogram, electroencephalogram, etc.), radiological explorations (RMN, TAC, conventional radiology, etc.), and direct
  observation, among others.
- Clinical Context: This is related to the variety of situations or states of an
  assistance act, like a revision consultation, a postsurgical consultation,
  an admission, an emergency assistance or a ward checkup. These contexts are obviously classified according to the medical speciality and, in
  some cases, to sub-speciality and process.
- Assistance act: It determines the origin of the assistance procedure
  (admission, consultation, emergency, etc.).
- Agent: This class is used to define the kind of professional that is involved in the act, locating him/her in the corresponding service and
  professional category (doctor, nurse, assistant, etc.).
- Archetype: We use the internally defined archetypes and those other
  defined by the different research groups working on the interoperability
  of the EHR [31].

- Application: This class captures the variety of functional applications from different providers and the specific tools, that the clinical workstations entail and must be integrated. These applications are complemented in the system ARCHINET by means of its own and specific functionalities, with a logistical or departmental character. Some examples are the application of medicine and unidosis management, the application of analytical requests management, or the emergency monitoring. Some of these applications may derive clinical data towards the EHR.
- ICD-10 Hierarchy: This reproduces the class structure in this international classification [24]. We have chosen this classification since it is the most habitually used in the hospital for diagnosis encoding.
- Data Model: This is a class of internal use for the procedures of computing agents. Its aim is to reproduce the data model starting from its logical modelling and down to reach the physical Data Base model of the EHR. This is how the tables stored in the data base are described. The instances of this class are each of the individual data (columns). The hierarchy of this class shows the typology of these structures: movement tables, primary tables, history tables, etc.

2.5.3 Properties in the Ontology

Regarding the properties in the ontology, their purpose is to create sets of restrictions based on the taxonomical relation between classes, in such a way that each possible entry in a EHR has a semantic map to contextualize its use, and hence its relations to other elements in the EHR. This way, as an example, the entry "anaesthesia type" belongs to the document "anaesthesia sheet", is a data type of restricted values, and is part of the context "intrasurgical information" and of the assistance procedure "Anaesthesia". In addition, it is characteristic of the assistance acts "admission", "emergency" and "surgical day hospital". Its agent profile is "anaesthetist doctor" and it is considered as related to the archetype "anaesthetic report". To allow the creation of these "semantic contextual maps" in the ontology we have defined complex relations between classes and their corresponding attributes and restrictions.

The process to create these relations is quite complex. However, it is easier using the information stored in the EHR Base and in the Access Base, and also referencing the data model itself. Most of them can even be automatically generated. In the ontology we have also included implicit properties for concrete classes like the "character of a document" (confidential or open), or

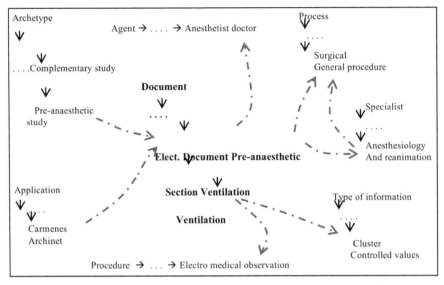

Figure 2.5 OWL based relationships between classes and their properties.

the "type of document" (general, of speciality, of process, logistical, etc.). Finally, we must remark that the definition of the ontology is not a closed topic, but a continuous process that, depending on the experiments, we expand or modify.

2.6 Ontology Applications

The creation of the ontology provides a knowledge base formalized with structures that the computing procedures can use to answer the query processes performed on the EHRs [20], and opens the possibility of using new accessibility models to the EHR. Concretely, it makes possible the conceptual accessibility of the data in the EHR, which opens the path towards the interoperability between EHR systems, since it provides the system with the ability to semantically interpret the clinical data retrieval processes. In addition, it sets the basis for the following uses of the information and the system:

- *Contextual use:* to allow the doctors to have the information really needed for the assistance activity in which he/she is involved, acceding just to a determined context. This way superfluous or not pertinent information items are avoided, as well as complex accesses with the navigation systems.

- *Restricted navigation:* used in the cases where only some concrete information items are needed, avoiding the unuseful navigation through acts, processes and documents with no interest to the search purposes.
- *Limited navigation for mobile devices:* The navigation through the contents of the EHR is quite difficult in mobile devices, since their screens set a very limited representation capability, especially for complex menus. In this case, the information presented can be initially focused according to a given work environment, like the medical speciality, the assistance act to be performed, the process or the assistance procedure. All of them set an environment to which the system can give a response depending on the information relevant to it.
- *Ontology navigability:* Traditionally there have been discrepancies regarding the different ways to show the documental organization of the EHR. Some times it is necessary to organize them according to assistance acts, whereas in other cases the organization according to processes is preferred. In our case the user can choose, with the scheme of classes that the ontology provides, allowing him/her to design of his/her own navigation model.
- *Interoperability:* This is easier to reach with the Knowledge Base provided, making possible the understanding with other formalized models, especially with the Reference and Archetypes models defined in the ISO 13606.
- *Access according to the semantic valuation:* This makes the direct access to elements contained in the EHR possible, using the terminology in the ontology.

As a summary, the ontology conceptualizes our model of EHR, opening the access to a great variety of opportunities to develop computing procedures to make easier the use, control and availability if the EHR.

To our best knowledge, there are some proposals of ontologies for contextualized access in others fields (e.g. e-Goverment, business context [9]) but none for EHR access so a comparison with our proposal is not possible.

2.7 Conclusions

In this chapter we have proposed a new way to solve the interoperability based on the standard ISO 16303 and ontologies. The result is an ontology that integrates the normal use of the system with the standard so the interoperability is considered inside the structure and not as an interface. To do so we have presented a semantic conceptualization model for an EHR system, that offers a number of utilities towards three purposes: the interoperatibility, the accessibility and the mobility. The proposed design can be generally and widely applied, independently of the document structure, the technological support or the development degree. The ontology provides a formalized knowledge base that allows the development of computing procedures with several purposes, from analytical to the accessibility, opening the path to new alternatives to the traditional navigation and access procedures to the EHRs.

The use of the ontology inside the EHR system allow us to introduce new functionality not only for interoperability but for other patient needs, such as the personal EHR systems.

References

[1] K. Abrams, S. Schneider, and R. Scichilone. Managing terminology assets in electronic health records. *Studies in Health Technology and Informatics*, 143:507–512, 2009.

[2] A. Balas and A. Al Sanousi. Interoperable electronic patient records for health care improvement. *Studies in Health Technology and Informatics*, 150:19–23, 2009.

[3] T. Beale. Archetypes: Constraint-based domain models for future-proof information systems. In *OOPSLA 2002 Workshop on Behavioural Semantics*. Citeseer, 2002.

[4] V. Bicer, O. Kilic, A. Dogac, and G.B. Laleci. Archetype-based semantic interoperability of web service messages in the health care domain. *International Journal on Semantic Web and Information Systems*, 1(4):1, 2005.

[5] J. Bisbal and D. Berry. An analysis framework for electronic health record systems. *Methods of Information in Medicine*, 48(1), 2009.

[6] B. Blobel, K. Engel, and P. Pharow. HL7 version 3 compared to advanced architecture standards. *Methods Inf. Med.*, 45:343–353, 2006.

[7] B. Blobel and P. Pharow. Analysis and evaluation of EHR approaches. *Methods of Information in Medicine*, 48(2):170–177, 2009.

[8] A. Brass, D. Moner, C. Hildebrand, and M. Robles. Standardized and flexible health data management with an archetype driven EHR system (EHRFlex). *Studies in Health Technology and Informatics*, 155:212, 2010.

[9] Hamdi Chaker, Max Chevalier, Chantal Soule-Dupuy, and Andre Tricot. Improving information retieval by modelling business context. In *Proceedings of Third International Conference on Advances in Human-Oriented and Personalized Mechanisms, Technologies and Services (CENTRIC)*, pages 117–122, 2010.

[10] R. Chen, G.O. Klein, E. Sundvall, D. Karlsson, and H. Åhlfeldt. Archetype-based conversion of ehr content models: Pilot experience with a regional ehr system. *BMC Medical Informatics and Decision Making*, 9(1):33, 2009.

[11] I. Cho, J. Kim, J.H. Kim, H.Y. Kim, and Y. Kim. Design and implementation of a standards-based interoperable clinical decision support architecture in the context of the Korean EHR. *International Journal of Medical Informatics*, 79(9):611–622, 2010.

[12] Citrix Systems. http://www.citrix.com.

[13] HL7 data exchange standard. http://www.hl7.org, 2011.

[14] S. Garde, S. Heard, and E.J.S. Hovenga. Archetypes in electronic health records: Making the case and showing the path for domain knowledge governance. *HIC 2005 and HINZ 2005: Proceedings*, page 267, 2005.

[15] C. Golbreich, O. Dameron, B. Gibaud, and A. Burgun. Web ontology language requirements w.r.t expressiveness of taxonomy and axioms in medicine. In *Proceedings of 2nd International Semantic Web Conference*, 2003.

[16] DICOM image storage standard. http://www.medical.nema.org, 2011.

[17] ISO-13606. *ISO 13606: Electronic health record communication*, 2008.

[18] P. Lambrix, M. Habbouche, and M. Perez. Evaluation of ontology development tools for bioinformatics. *Bioinformatic*, 19(12):1564–1571, 2003.

[19] J.A. Maldonado, D. Moner, and et al. Semantic upgrade and normalization of existing EHR extracts. In *Proceedings of the 30th Annual International Conference EMBC*, 2008.

[20] J.. Mei and E.P. Bontas. Reasoning paradigms for SWRL-enabled ontologies. In *Proceedings of Protégé with Rules Workshop, 8th International Protg Conference*, Madrid, Spain, 2005.

[21] M. Amparo Vila Miranda Miguel Prados de Reyes, M. Carmen Peña Yáñez and M. Belen Prados Suárez. Generation and use of one ontology for intelligent information retrieval from electronic record histories. 2006.

[22] Open-EHR Open Electronical Health Records. http://www.openehr.org, 2011.

[23] Oracle. http://www.oracle.com.

[24] World Health Organization. International classification of diseases (ICD).

[25] M. Prados, M.C. Peña M.C., M.B. Prados, and J.M. Garrido. *Gestión de conocimiento en el ámbito hospitalario*. EASP (Escuela Andaluza de Salud Pública), 2004.

[26] M. Prados and M.C. Peña. *Sistemas de Informacin hospitalarios. Organizacin y gestin de Proyectos*. EASP, 2003.

[27] M. Prados, M.C. Peña, B. Prados, B. Martinez, J.C. Ortigosa, and A.C. Delgado. Electronical health record (EHR) representation through ontology. mobility, accesibility and interoperability usefulness. 2010.

[28] B. Prados-Suárez, E. Gonzlez Revuelta, G. Carmona Martínez C. Peña Yáñez, and C. Molina Fernández. Ontology based semantic representation of the reports and results in a hospital information system. In *Proceedings of the ICEIS 2008*, pages 300–306, 2008.

[29] M.R. Santos, M.P. Bax, and D. Kalra. Building a logical EHR architecture based on ISO 13606 standard and semantic web technologies. *Studies in Health Technology and Informatics*, 160:161, 2010.

[30] P. Serrano, D. Moner, T. Sebastián, J.A. Maldonado, R. Navalón, M. Robles, and Á. Gómez. Utilidad de los arquetipos ISO 13606 para representar modelos clínicos detallados. *RevistaeSalud. Com.*, 5(18), 2009.

[31] E. Sundvall, M. Nystrom, M. Forss, R. Chen, H. Petersson, and H. Ahlfeldt. Graphical overview and navigation of electronic health records in a prototyping environment using Google Earth and open EHR archetypes. *Studies in Health Technology and Informatics*, 129(2):1043, 2007.

3

Semantic Technologies for the Modelling of Human Behaviour from a Psychosocial View

Ana Belén Sánchez-Calzón, Teresa Meneu and Vicente Traver

ITACA-TSB, Universidad Politécnica de Valencia, Camino de Vera S/N, 46022 Valencia, Spain; e-mail: {absanch, tmeneu, vtraver}@itaca.upv.es

Abstract

Maintenance of responsible attitudes to a healthy lifestyle is highly beneficial. Health is a resource that gives people the ability to freely manage their environment. Health promotion supports people to control and improve their own health, underlining the role of individuals in its definition. Nevertheless, people need motivation and enough support to achieve changes in health-related behaviour. To design and develop a model that properly predicts and anticipates human behaviour represents a difficult task, in the sense that the challenge of representing the behaviour includes such different areas as motivating the effects of making decision, modelling the cognitive processes that take place in making decision, or stimulating the perception of motor skills. The examination of an integrated methodology, according to the main schools of thought of human behaviour, will facilitate the identification of the preeminent technologies focused on the automatic behaviour modelling based on pattern recognition research.

Keywords: human behaviour modelling, semantic technologies, psychosocial.

S.F. Pileggi and C. Fernandez-Llatas (Eds.), Semantic Interoperability: Issues, Solutions, and Challenges, 49–67.

3.1 Introduction

The study of the behaviour motivators represents one of the most complex current challenges in different research fields. Human behaviour is determined by biological, psychological and socio-cultural factors. There are multiple and varied factors that make it difficult to find and point out the conditions that influence behaviour, as well as the different ways to represent that knowledge. The basis for human behaviour has been treated from different research fields throughout history, especially from Psychology and Sociology [14, 17]. Health promotion strategy is a discipline in which the behaviour discovery has more applications. The promotion of healthy habits has many benefits at a personal and socio-economic level. Proper health care of individuals is important, and also the inherent amount of money that can be saved in treatments for illnesses directly related with unhealthy lifestyles. Year by year, politics and governments invest a significant part of their budget in promoting healthy habits. Nevertheless, despite this investment, it is very difficult to know what the precise impact of their activities is. In the literature, some emerging technologies in the automatic human behaviour modelling based on pattern recognition research could be used to study the abilities of health promotion techniques applied by governments and healthcare professionals. Complex Event Processing [31] techniques allow the processing of events and notice complex patterns among multiple streams of event data. Plan Recognition Models [22] allow the alignment of human habits and routines, in order to detect unhealthy situations and evaluate the submission to health promotion campaigns. Process Mining techniques (A.K.A., Workflow Mining) [9] allow the detection of changes in the behaviour, identifying if health promotion strategies have enough impact on them. Nonetheless, the totality of different theories and schools of thought make it difficult for those techniques to determine what factors should be taken into account. Although the different psychological and sociological research lines seem to be theoretically complementary, an effort to create a common framework is needed, in order to condense the main theories using ontologies to describe the concepts, also to provide a mechanism that interoperate among them semantically. Behavioural theories and frameworks provide a large body of knowledge for understanding what influences behaviour. An extensive analysis of principal theories and schools of thought was done to reveal how people can be profiled for their health conduct and motivational factors, and how this information can be used to select the recommended interventions for changing identified unhealthy behaviours. In this chapter, a review about the factors involved on human behaviour modelling, as well as the most important theories on people health behaviour is done, in order to approach to a unified health behaviour model.

3.2 Factors Involved on Human Behaviour Modelling

The multitude of factors involved in the initiation, maintenance and accomplishment of a given action makes the study of human behaviour a highly complex task. One of the main objectives of Social Psychology is the study of the diverse processes that integrate and determine behaviour. Social scientists know that human behaviour is directly observable, but not the psychological processes that happen before, during or after the execution of such behaviour. Therefore, it is necessary to assume that there are several factors involved in the production and manifestation of a specific behaviour. Social psychology seeks to understand and provide explanations of behaviour, and then, be able to predict it before it happens. The main point is to anticipate events, knowing the behaviour likely to occur, and also know what conditions lead to it, both individual and social, in order to analyze and explain it through the incorporation of semantic technologies allowing us to understand, relate, interpret and classify the implicit knowledge in digital contents. The semantic interoperability represents the capacity created by the application of such technologies. In general, human behaviour can be predicted and modeled taking into account biological, psychological and social factors.

3.2.1 Psychobiological Factors

Psychobiology focuses on the study of human behaviour as a biological property that allows the organism to establish an active and adaptive reaction to the environment. Thus, the objective of psychobiology is studying biological systems and processes involved in behaviour, and finding out how natural selection is shaping these processes and the behaviour, contributing to the development of behavioural models that can be seen in different animal species, including humans [23]. From the point of view of science, the structural, physiological and behavioural characteristics of an individual are the result of two factors:

- The *phylogenetic issue* concerns the evolutionary history of the human species, a factor reflected in the individual genetic information, through which the adaptive advances of the species are transmitted from generation to generation, as well as the general characteristics.

- The *ontogenetic issue* relates to the circumstances through which the phylogenetic factor is modulated by the internal and external environment since the beginning of the individual's life [23].

The *phylogenetic issue* determines the general characteristics such as the type of sensory organs, the regulatory systems of the internal environment, locomotion systems, etc. All these features will determine what stimuli the individual can collect and what kind of response this will lead to. In addition to these general characteristics, according to psychobiology, there are also variations among individuals, caused by genetic variability present in all species and by the interaction between genetic and environmental factors.

Therefore, from the point of view of this scientific development, understanding, anticipating and modelling human behaviour requires a deep knowledge and analysis of biological characteristics of the individual, examining how these characteristics are determined by genes, what mechanisms influenced the genetic information during evolution and, finally, what the characteristics are of the neuroendocrine system that allow regulating the dynamic relationship between the individual and her environment, that is, to manifest a particular behaviour [6].

3.2.2 Psychological Factors

Regarding the strictly psychological factors, psychology focuses on the scientific study of behaviour and experience, how people feel, think, understand, learn, interpret and know in order to adapt to their environment. Within the field of psychology there are many schools of thought and currents of opinion that provide particular definitions of behaviour and the prediction of behaviour. Psychobiology, described in the previous section, is one of those currents and, in addition, we also briefly introduce other schools of thought:

- *Behaviourism* is characterised by gathering facts about the observed behaviour, systematically organizing and developing theories to describe it, without paying attention to any explanation. It is a theory based on scientific method, focusing on the conditions which determine the behaviour, following a cause-effect scheme, which allows predicting the behaviour and the ability to mediate the conditions that promote it. In this regard, computer technology is one of the common methodologies used to develop predictive models of behaviour.
- *Cognitivism* focuses on the study of mental life processes concerning knowledge, i.e. the mechanisms of development of knowledge.

The cognitive issue refers to the act of knowing, as in storing, recognising, understanding, organising and using the information received through the sensory organs, in short: our senses. Nowadays, it also relies on computer models to achieve explanation and understanding of different cognitive processes. Basic psychological functions studied by cognitivism are:

- *Attention*, the mechanism in which the individual is made aware of certain contents of his mind over others.
- *Perception*, the ways in which the individual responds to stimuli gathered by the sensory organs (senses). They are the processes of immediate experience.
- *Memory*, a mechanism that allows the individual to code, organise, store and recover information.
- *Thought*, the set of processes that allow the organism to develop the information collected or stored in memory.
- *Language*, the representative system of signs and rules that represent a symbolic form of communication between people.
- *Learning*, the process in which individuals acquire new skills, knowledge, values, beliefs and behaviours, resulting from the study, analysis, observation, reasoning, instruction and experience [18].

- *Social Psychology* is an interdisciplinary field between psychology and sociology, it studies how individuals, interacting in society, do things together, how they use language and develop personal and social identity, including how those identities are shaped by various factors such as ethnicity, gender, nationality or age. This science studies social phenomena, trying to set parameters that govern the coexistence. It investigates social organisations, looking for patterns of behaviour of individuals within groups, studying the roles and situations that influence and determine how they act [18].

Starting from this brief sample of the different lines of thought and existing research in psychology, one may distinguish between emotional characteristics (moods, emotions, etc.) and cognitive characteristics (beliefs, expectations, etc.) of individuals. Most research focused on the study of factors that can predict human behaviour points out the cognitive aspects, especially attitudes, above other equally important factors involved in the maintenance of a behaviour, namely the biological, social and emotional factors, which are inconveniently relegated, in many cases, to a second plane. Attitudes are the positions that are taken about an idea or action, and the tendency to act in a certain way about an object or situation. It is the permanent predisposition to react in a certain way in a given situation.

3.2.3 Social Factors

Regarding the social factors (networks, norms, conventions, etc.), they act by facilitating or inhibiting the expression of a given behaviour. The main goal of human behaviour is to relate to other people properly, establishing links between them, so the human development requires the presence of reference models and reinforcement systems that are provided by the others. Through society, the individual adapts to the environment, in which they are involved in two types of factors: material factors surrounding the individual; and human factors, consisting of those persons who are part of the individual's life, who influence him/her directly or indirectly by providing him/her targets to achieve, and thus determining the perception of the individual. Through socialisation, it activates a set of processes through which individuals internalise the social norms that shape their personal identity [14]. This is a process of internalisation that means their adaptation to a concrete society and culture, a process that occurs at three levels:

- Biological and psychomotor level (wake-sleep, food, clothing, etc.).
- Emotional level (forming the emotional life in the environment in which the individual develops).
- Thinking level (grasping and adopting values, and adding categories of thoughts).

The existence of wide and diverse information about research focused on modelling human behaviour involves, at a level of research, the challenge to find solutions to integrate a whole range of information systems and models of intervention. The design, development and implementation of semantic technologies in the context of the human behaviour study can optimise the search processes, identify relationships between resources, exchanging messages between systems, allow the customisation of interest and specific features, and interpret the meaning and data context, obtaining the contents of data fields using a standard common code.

In the probability of executing a particular behaviour various factors should be taken into account such as, for example, the type of behaviour, the goal intended by behaviour, the place where the behaviour is performed, or the time when the behaviour occurs. There are many factors that precede and explain human behaviour. In recent years many models were developed focused on the field of healthy behaviour, disease prevention and health promotion. These are models that differ in terms of theoretical perspective and category of behaviour; however, they are useful in the present study because they contain similar kinds of variables, and they can be integrated additionally. Below are the most relevant.

3.3 Psychosocial Theories in Human Behaviour Modelling

Based on the behaviour factors previously described, according to current criteria of professional scientific research, a group of theories implemented are shown in order to explain how these factors influence in health motivation and the resultant behaviour. The most important theories are the Health Belief Model, the Theory of Reasoned Action, the Social Action Theory, and the Self-Efficacy Theory.

3.3.1 Health Belief Model

The theoretical development was first formulated by Hochbaum [12], and successively extended by Rosenstock [27], applied to the explanation and prediction of a wide range of health behaviours. The scheme is the explanation of behaviour by focusing on psychosocial variables, from a group of psychological theories such as Field Theory by Lewin [16, 17], Expected Value Theory, and the Theories of Decision Making under Uncertainty. The initial hypothesis of this model is that an individual does perform healthy behaviour in the following conditions:

- If he/she has minimum levels of relevant motivation and information toward health.
- If he does not see himself as potentially vulnerable.
- If he/she perceives the disease (or risk) as threatening.
- If he/she is convinced of the effectiveness of the intervention.
- If he/she find little difficulty in the practical action of healthy behaviour [27].

According to the Health Belief Model, the probability of displaying a healthy behaviour depends on the subjective state of intention that the individual has to do so. The intention is determined by the threat of the disease in question (according to the belief of the individual). The perceived threat is determined by the perceived susceptibility to the disease, the perceived severity of the consequences of suffering the disease, and keys to trigger action to proper health behaviour. These keys may come from internal sources (symptoms and so on) or external (interactions with others, media, etc.) [5].

Secondly, the probability that an individual develops and maintains a healthy behaviour depends on an evaluation that makes such behaviour in terms of viability and usefulness, disputed with the perceptions of the physical environment, economic costs and other barriers involved in the proposed action. The perceived costs have been measured in different ways in different research: in terms of security on the effectiveness of the treatment prescribed, on the patient satisfaction about the communication with the health professional, on patient satisfaction about matters such as the way in which the health organisation provides medical care and so on.

Regarding the threat perception, it is triggered by a main event starting the process of making wholesome behaviour. The Health Belief Model is based on the premise that socio-demographic, structural and individual factors may determine health behaviours. However, the model considers that these variables act through their effects on individual health beliefs, and are not direct causes of healthy actions [5].

3.3.2 Theory of Reasoned Action

Fishbein and Ajzen [19] established the Theory of Reasoned Action, developed later as the Theory of Planned Action [1, 2], which tries to explain those behaviours under conscious control of individuals, from different determinants that precede and explain them. According to the authors, the immediate determinant of behaviour is the intention to do it. Behavioural intention has two precursors that explain it: an individual precursor, i.e., the attitude about the behaviour; and also another social and collective which refers to the socio-cultural context of the person, known as the subjective norm [19]. Both attitude and subjective norm are determined by other factors that precede them, helping to better understand the behaviour. Attitude is determined by each of the beliefs that person has to an object, and the assessment based on these beliefs. This valuation represents the affective component of attitude, determining the motivation and the strength of behavioural intention.

According to the theoretical construct, beliefs vary according to their source, and they can be formed from different processes:

- *Direct experience regarding the attitude object*, through which information is collected on the characteristics of the object (one person, thing, etc.). The attitudes shaped from this process have greater strength, being more resistant to changes.

- *Indirect experience regarding the attitude object*, which brings the same features of that object by the similarity to other objects with which it has had previous direct experience. Such beliefs are named inferential.
- *The information collected from other* (media, family, friends and so on). Information is assumed as real, unless it contradicts the beliefs formed from the direct or indirect experience [19].

Regarding the subjective norm, it is determined by the perception of the beliefs that others individuals have about the behaviour the individual must perform and, moreover, it is also determined by the individual's motivation to meet the expectations that others have about him/her. It is a differential process of beliefs formation through which each of the beliefs have a particular weight and value, according to each person and the attitude object. The knowledge about the specific beliefs, or what others think of each of the specific behaviours (for instance physical exercise, healthy eating, etc.) will exert some influence over the intention to perform or not a given behaviour, depending on the motivation.

Nevertheless, not all behaviours are intentionally controlled by the individual, because there are many situations in which happenstance may occur, or in which skills that could interfere with the intention of performing behaviour are required. It is necessary to introduce a third determinant of behavioural intention, perceived control [1, 2]. Although the individual has a favourable attitude toward behaviour, the probability of carrying it out will depend, mainly, on the perception of control by the individual about his/her behaviour. According to the theory, people may have a favourable attitude toward healthcare, for instance to give up smoking. But, if someone perceives a low ability to quit smoking due to little faith in his/her capacity, or because he/she believes that behaviour of others can interfere with his/her decision to quit, this healthy behaviour will not be performed. The perception of control is a factor consisting of internal variables (perceived ability, skill of action, etc.), and external variables (opportunity to action, obstacles, time, cooperation and so on). This is a determinant that helps to improve predicting and modelling of the behaviour [2].

The Theory of Reasoned Action does not give weight enough to relevant factors such as attitudes towards goals, personality features, sociodemographic variables, social role and so on. For this model, these factors are external variables that can influence behaviour, but there is no necessary relationship between them.

3.3.3 The Social Action Theory

This model proposed by Ewart [8] presents the person as a self-regulatory system that is actively trying to achieve objectives, and also as a feedback system consisting of a set of successively arranged stages:

- First input stage which represents stimulus field and sets the targets.
- Second output stage, also known as production of response, which provides the plans, the selection and development of responses.
- Third stage of supervision, which involves the consideration of the consequences of action in relation to the initial set of objectives.

The theory underlines the role of social context in the development and maintenance of healthy habits. It provides the causal structure linking the self-change processes in interpersonal environments. And it also specifies the social and environmental influences that make possible personal changes.

According to the Social Action Theory, preventive interventions include the creation of protective habits in the form of routine behavioural sequences that reduce personal risk factors. The actions are guided by their consequences in a control loop, and variations in the results generate compensatory behavioural regulations. The result is not a steady state-action, but constantly fluctuating. The starting point for intervention is an analysis of the relations between the harmful/beneficial behaviours to health and their effects experienced. This is an analysis that allows extracting those aspects in which behaviours are more accessible to prevention; it also allows extracting effective processes for the design of new schemes that promote health [8].

The Social Action Theory remarks that personal action schemes are socially interconnected to schemes of close social environments (friends, family, peers, etc.), so they raise significant obstacles to long-term changes. It is necessary to extend the concept of state-action focused individually to include social interdependence as a determinant of a modification in behaviour. Close social relations mean that the patterns of action of each person are interconnected, increasing the likelihood of someone trying to modify a routine, influencing and conditioning routines of others.

This model involves the existence of mechanisms that enable people to make transitions from old states-action to other new, causing a change. According to the theory [8], attitudes and reinforcements do not determine and cause behaviour. People motivate themselves by taking into account possible outcomes, assessing their capabilities, and creating goals that guide the solution of various problems. Health related behaviours are included, according to the Theory of Social Action, in a set of norms focused toward some important goal. This set of schemas is composed by the individual's personal projects, refers to basic tasks (such as getting social influence, acquiring material resources, be accepted by others, etc.), and finally it affects protective rules of behaviour so people can generate aimed objectives to evaluate their own conduct.

The schemas represent systematised sets of behaviours that focus our attention to specific aspects of situations and scenarios. They drive the encoding of experience in memory over the time. And finally they provide routines to do tasks. These ways of knowledge integrate capabilities which allow considering alternative goals, also to create new action strategies. Capacities are the mechanisms through which physical and social contexts affect to self-regulatory behaviours. Cognitive control schemes influence the varieties of behaviour, increasing trust on the individual's ability to maintain or change behaviour. The context in which people live is a tool that adjusts the personal capacities and the social relations, affecting the formation of goals, the consideration of opportunities for action, and the design of relevant health strategies. The physical characteristics of the environment determine the access to material resources, and they also influence in the behavioural strategies. Social relationships include a set of benefits, expectations and obligations that likewise influence on the objectives aimed. In addition, social relationships provide behaviour models that facilitate or inhibit the action guidelines [25].

3.3.4 Self-Efficacy Theory

Bandura [3] developed the Self-Efficacy Theory to explain human behaviour and those factors involved in motivation. Self-efficacy can be defined as the evaluation of one's own personal abilities in view of the possibility of action. There are different processes that facilitate the configuration and development of self-efficacy. They are elements that describe internal and external characteristics of the people, and also help to shape the beliefs that they have about what they are able or unable to do [4]:

- *Direct experience* is the main source of formation regarding the self-concept an individual has. In this sense, the consequences experienced after executing the behaviour make it possible to report about one's ability to carry out the same behaviour, and if is possible to control the situational variables in which the behaviour has to be done. Hence, the experience and results achieved contribute to the formation of self-concept, and also contribute to develop the personal self-worth feeling, both required to deal in different situations.

- *Learning by observation* is also a source of valuable information. To analyze the consequences that any action can produce in another individual, the execution of such action can lead either to inhibit or to promote one's own action, depending on positive or negative assessment of the observed consequences in other individuals. It is a process of social comparison in which the individual shape his/her perception of his/her own capabilities to deal with different situations.

- *Persuasive messages*: The attempts of people in the environment to encourage the person to take a specific action can provide security and support, in order to induce the behaviour. However, persuasion is a weaker resource than previous ones, and it will vary depending on different variables such as, for instance, certain features related to the personality of the individual who tries to persuade, the credibility that individual has to pass on, or his/her ability to achieve that another individual execute the action.

- *Physiological activation* may also be an important modulator of the capabilities that people believe they have. It can influence the process of self-valuation. According to Bandura [4], the information provided by the psycho-physiological activation influences the perceived effectiveness through assessment processes. Therefore, when such activation takes place in the view of the possibility of executing a behaviour, the individual evaluates different factors, stressing the sources that cause the activation of behaviour, the intensity of activation, the conditions in which that activation happens, and the way in which the activation influences over the efficacy. When physiological activation is successful after the execution of a given behaviour on previous experience, this activation is considered by individuals as a facilitator of the action. In opposition, when it is unsuccessful, it is considered as harmful or inhibitor of the action [4].

People increasingly develop their self-efficacy perception based on:

- Executed behaviours.
- Assumed explanation in their environment.

- The reinforcement that other people around them respect the behaviour they have conducted.

3.4 Review of Psychosocial Intervention Research

In scientific literature, there are multiple examples of how psychosocial factors are taken into account to work on the motivation and other conditions of health behaviour [7,10,13,28]. The complexity of human behaviour makes difficult to fully measure each of the factors involved in every moment and situation. Nevertheless, the more variables can be defined within a construct, the more information can be obtained from the different aspects that interfere with the execution and, consequently, the more accurate prediction can be performed, as well as the establishment of a helping plan to improve the execution of such behaviour. Therefore, the comparison of theories and models is intended to improve both the content of the measuring instruments and the prediction of behaviour.

Until the late 1980s, models as previously reviewed were mainly used in market research and vote intentions studies. However, from that decade onwards, the application in predicting the behaviour in areas such as health care increased. Specifically, many studies have focused on predicting different preventive behaviours and the maintenance of medical advice. Rodgers and Brawley [24] used the Theory of Planned Action and the Self-Efficacy Theory to study and predict if individuals would continue or cease the participation in a weight control program, in order to improve health and prevent disease. This research focused on developing an educative program about nutritional aspects and promoting self-monitoring of diet and exercise, using the components of these two theories.

Sparks and Guthrie [29, 30] conducted research focused on attitudes and intentions of people towards the fulfillment of a diet low in animal fat, in order to prevent cardiovascular diseases, using the components of Theory of Planned Action. The study is based on the implementation of a follow-up questionnaire consists of a set of items that include several aspects related to attitude, subjective norm and control perception.

Hazavehei et al. [11] analyzed the effectiveness of a health education intervention based on the Health Belief Model, in order to reduce the risk of osteoporosis disease in female adolescents. Based on the premise that it is important to start the primary prevention at an early age, even the most favourable way for set up this preventive behaviour in youth is difficult to define. The population chosen was female students from the middle schools of the city of Garmdsar, Iran, and the results of the study supported the practicability of a health education program based on Health Belief Model to induce behaviour change for osteoporosis prevention.

These are examples of different studies that have tested the predictive utility of these explanatory and modelling behaviour theories. Scientific research has established and concluded that attitudes and perception of control are the best predictors of health behaviours, in that they are the subjective norm controlling the variable influences in the intention to act.

Attitudes play a very important role in predicting and modelling human behaviour, while belief (cognitive variables), and evaluations of them (affective component) influence the decision of an individual to behave in a certain way. Rosenberg and Hovland [26] state that attitudinal systems can be explained from a series of correlated dimensions: The behaviour, the intention to carry out this behaviour, cognitions, and affective responses. These dimensions influence the affective value to healthy behaviour. Accordingly, the attitude is formed by all these elements, and it determines the following behaviour.

If attitudes determine human behaviour, to learn about the factors that influence attitudes can help us to predict and model the behaviour in many contexts. Attitudes are durable evaluations of different aspects of the social world, assessments that are stored in memory [20]. Attitudes often work as schemas or cognitive frameworks that organise information on specific concepts, situations or events, influencing the way in which people process the social information. Attitudes represent an association between an object and a given situation. It is a lasting organisation of beliefs and cognitions in general, containing an emotional weight in favour of or against a defined purpose, which predisposes a rational action in relation to the conditions and affections connected to object [1]. There are three main basic components of attitudes: a cognitive component, which represents the knowledge an individual has of an object. The affective component, which is the feeling that the object causes to the individual. And a behavioural component, which includes trends and intentions towards the object, as well as the actions directed towards it [26].

3.5 To a Unified Semantical Model to Represent the Psychology of Health Behaviour

It is possible to think of a proper integration of the different models and variables reviewed in this chapter, i.e., a complete integration of the diverse

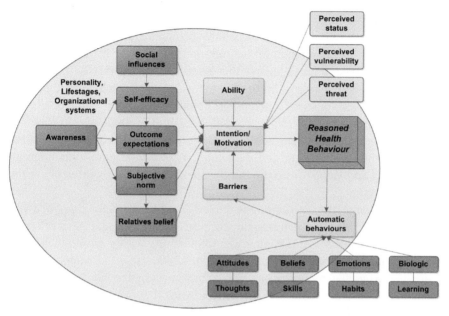

Figure 3.1 General architecture of the integrated semantic model.

explanatory models in a functional analysis of preventive and healthy behaviour. The theories previously described have a conceptual structure that can be applied to any matter related to health. Those models represent useful tools for numerous investigations that help to increase scientific knowledge about the health-related lifestyles. Secondly, they may help to organise and explain correctly the numerous observations accumulated. And finally, they may provide a guideline to action, in order to propose suitable interventions for modelling behaviour [25]. These are models that can help to realise questions, then to transform into methodological procedures to ensure the validity and reliability of the analysis and observations that are made.

The classic and most relevant theoretical models of behavioural determinants may overlap and complement each other, providing an integrated model which covers the key determinants of rational behaviour and behaviour change. In Figure 3.1, a general approach to a complementary model of theories is presented. In the schema, the different models are interconnected in a collaborative way according to their central hypothesis, in order to progress in the analysis of the psychosocial determinants of health behaviour.

For health behaviour modification to have effect, people need the *abilities* required by the change process, the *motivation* to address the change process, and be/feel free from any *barriers* inhibiting the adjustment. In order to the intention to develop a given healthy behaviour, the individual, in first place, has to be *aware* enough of the importance to change his/her behaviour. Secondly, the person has to believe that he/she can succeed with that change (*self-efficacy*), which means, he/she has to feel that he has encouraging *social influences*, and also positive *expectations* about the outcomes of the behaviour modification. In addition, the individual takes into account two factors: the *subjective norm*, which represents the perceived social pressure to involve or not in a given behaviour; and the *beliefs* the *relatives* have, which are all the opinions and values from the family members and peers influencing the intention to act.

In the schema, the target point is the *reasoned health behaviour*. The aim is to point out the main determinants of such behaviour, and their relationships. There are *automatic behaviours* founded on lasting *attitudes, beliefs, skills, emotions, habits, thoughts* and so on, which can either support or obstruct the behaviour modification. People have personal values, a *personality* defined through their successive *life stages*, inside the social environment, i.e., the *organisational system*. These aspects set the background for the decisions, defining largely the determinants of behaviour.

The degree of *perceived status, perceived vulnerability*, and *perceived threat* is directly connected to the motivation of the individual to engage or not a healthy behaviour. They are preceding variables of behaviour.

Consequently we must take into account the contextual variables, both physical and biological, such as social variables. The formal exemplification of the results through ontologies allows their use not only in the same research field but also in the others. Based on that unified model, and illustrating the subsequent knowledge of these theoretical approaches in semantically identified concepts, it would be possible to perform automated reasoning on the theories of health behaviour Psychosociology.

The benefits of personalised intervention methods are well founded in health promotion research [15, 21]. A careful approach focused on individual characteristics such as values, beliefs, abilities, needs or interests has a better chance to motivate in order to influence attitudes, and lead to a proper behaviour change.

3.6 Conclusions

Health habits are determined by early socialisation. Routines such as a healthy diet, taking care for oral hygiene, frequent physical activity, enough sleep, avoiding alcohol and tobacco, commit to a responsible lifestyle and so on, are established within the family group, mainly through imitating behaviour of parents and other relatives. On the other hand, the value systems of socio-cultural reference organisation have a strong influence on the lifestyle of people. Finally, social groups are a very important factor in the development, maintenance or conclusion of health habits. One of the social context elements related to health care, and therefore to lifestyles, is social support. Most of the programs of health promotion and disease prevention are developed on the basis of social support networks. Social support is a powerful factor to achieve compliance with different preventive strategies, by providing social validation and appropriate models of preventive behaviour, and also through the social reinforcement provided by significant others (family, peers, etc.).

The use of ontologies to allow the semantic interoperability between systems, which represent the several health behaviour models, will enable the creation of an integrated model that explains the entire health behaviour of people. According to this, scientists and experts will be able to share not only in theory but also their practical knowledge, which could be used by other research fields. These systems can offer a better knowledge that can be used in combination with emerging pattern recognition theories, in order to validate the use of health promotion practices.

References

[1] I. Ajzen. Attitudes, traits, and actions: Dispositional prediction of behavior in personality and social psychology. *Advances in Experimental Social Psychology*, 20(C):1–63, 1987.

[2] I. Ajzen. *Attitudes, Personality, and Behavior*. The Dorsey Press, 1988.

[3] A. Bandura. Self-efficacy: Toward a unifying theory of behavioral change. *Psychological Review*, 84(2):191–215, 1977.

[4] A. Bandura. Self-regulation of motivation and action through goal systems. In *Cognitive Perspectives on Emotion and Motivation*. Kluwer Academic Publishers, 1988.

[5] M. Becker and L. Maiman. Models of health-related behavior. In *Handbook of Health, Health Care and the Health Profession*. Free Press, 1982.

[6] M. Bunge and R. Ardila. *Philosophy of Psychology*. Siglo XXI, 2002.

[7] E. del Hoyo-Barbolla, R. Kukafka, M.T. Arredondo, and M. Ortega. A new perspective in the promotion of e-health. *Studies in health technology and informatics*, 124:404–412, 2006.

[8] C.K. Ewart. Social action theory for a public health psychology. *American Psychologist*, 46(9):931–946, 1991.

[9] Carlos Fernández-Llatas, Juan Pablo Lázaro, and Jose Miguel Benedí. Workflow mining application to ambient intelligence behavior modeling. In *Universal Access in Human-*

Computer Interaction, Lecture Notes in Computer Science, Vol. 5615, pages 160–167. Springer, 2009.

[10] David H. Gustafson, Michael G. Boyle, Bret R. Shaw, Andrew Isham, Fiona McTavish, Stephanie Richards, Christopher Schubert, Michael Levy, and Kim Johnson. An e-health solution for people with alcohol problems. *Alcohol Research & Health*, 33:327–337, 2011.

[11] S.M. Hazavehei, M. Taghdisi, and M.H. Saidi. Application of the health belief model for osteoporosis prevention among middle school girl students, Garmsar, Iran. *Education for Health (Abingdon, England)*, 20(1):23, 2007.

[12] G. Hochbaum. Public participation in medical screening programs: A sociopsychological study. Technical report, PHS publication No. 572. U.S. Government Printing Office, Washington DC, 1958.

[13] Y. Hu and S.S. Sundar. Effects of online health sources on credibility and behavioral intentions. *Communication Research*, 37(1):105–132, 2010.

[14] H. Kelman. La influencia social y los nexos entre el individuo y el sistema social; ms sobre los procesos de sumisin, identificacin e internalizacin. In *Estudios Basicos de Psicologia Social*. Ed. Hora., 1984.

[15] Paul Krebs, James O. Prochaska, and Joseph S. Rossi. Defining what works in tailoring: A meta-analysis of computer-tailored interventions for health behavior change. *Preventive Medicine*, 51(3-4):214–221, 2010.

[16] K. Lewin. Defining the 'field at a given time'. *Psychological Review*, 50(3):292–310, 1943.

[17] K. Lewin. Action research and minority problems. *Journal Social Issues*, 2(4):34–46, 1946.

[18] Lindesmith, Strauss A., and N. A. & Denzin. *Psicologa social. Madrid: Centro de Investigaciones Sociolgicas.* Siglo XXI de Espana Editores, S.A., 2006.

[19] M. Fishbeina and I. Ajzen. *Belief, Attitude, Intention and Behavior: An Introduction to Theory and Research*. Addison-Wesley, 1975.

[20] S. Moscovici. *Psicologa social: Influencia y cambio de actitudes. Individuos y grupos.* Paidos, 1985.

[21] Seth M. Noar, Nancy Grant Harrington, Stephanie K. Van Stee, and Rosalie Shemanski Aldrich. Tailored health communication to change lifestyle behaviors. *American Journal of Lifestyle Medicine*, 5(2):112–122, March/April 2011.

[22] Clifton Phua, Victor Siang-Fook Foo, Jit Biswas, Andrei Tolstikov, Aung-Phyo-Wai Aung, Jayachandran Maniyeri, Weimin Huang, Mon-Htwe That, Duangui Xu, and Alvin Kok-Weng Chu. 2-layer erroneous-plan recognition for dementia patients in smart homes. In *Proceedings of the 11th International Conference on e-Health Networking, Applications and Services*, Healthcom'09, Piscataway, NJ, USA, pages 21–28, IEEE Press, 2009.

[23] R. Frankel, T. Quill, and S. McDaniel. *The Biopsychosocial Approach: Past, Present, Future*. New York, 2003.

[24] W.M. Rodgers and L.R. Brawley. The influence of outcome expectancy and self-efficacy on the behavioral intentions of novice exercisers. *Journal of Applied Social Psychology*, 26(7):618–634, 1996.

[25] J. Rodriguez. *Psicologia Social de la Salud*. Sintesis Psicologia, 2001.

[26] M. Rosenberg and C. Hovland. Cognitive, affective and behavioral components of attitudes. In *Attitude, Organization and Change*. Yale University Press, 1960.

[27] I. Rosenstock. Historical origins of the health belief model. *Health Education Monographs*, 2, 1974.

[28] S. Chatterjee and A. Price. Healthy living with persuasive technologies: Framework, issues, and challenges. *Journal of the American Medical Informatics Association*, 16(2):171–178, 2009.

[29] P. Sparks and C.A. Guthrie. Self-identity and the theory of planned behavior: A useful addition or an unhelpful artifice? *Journal of Applied Social Psychology*, 28(15):1393–1410, 1998.

[30] P. Sparks, C.A. Guthrie, and R. Shepherd. The dimensional structure of the perceived behavioral control construct. *Journal of Applied Social Psychology*, 27(5):418–438, 1997.

[31] Segev Wasserkrug, Avigdor Gal, Opher Etzion, and Yulia Turchin. Complex event processing over uncertain data. In *Proceedings of the Second International Conference on Distributed Event-Based Systems (DEBS'08)*, New York, pages 253–264, ACM, 2008.

PART III

Services and Systems

4

Towards Semantic Interoperability in Industrial Production

Matthias Loskyll

*German Research Center for Artificial Intelligence (DFKI GmbH), Innovative
Factory Systems, Trippstadter Strasse 122, 67653 Kaiserslautern, Germany;
e-mail: matthias.loskyll@dfki.de*

Abstract

The Factory of Things describes the extension of the Internet of Things
specific to the production domain. The semantic description of products,
processes and plants depicts a basic module of this approach, through which
information can be filtered and services can be discovered and orchestrated
on demand. The usage of semantic technologies in the context of production
can make a valuable contribution towards managing the growing complexity, increasing the flexibility of production processes and establishing a
semantic interoperability among heterogenous systems. This chapter presents
an approach to build semantic plant models based on production-specific
knowledge sources, lexical knowledge and the usage of methods for semi-
automatic knowledge acquisition. The efficient development of the semantic
plant model based on a clearly defined methodology is meant as a first step
towards semantic interoperability in industrial production. We describe a case
study in an industry-related production plant, in which the developed semantic plant model serves as the basis for the efficient discovery and dynamic
orchestration of field device services to build adaptive production processes.

Keywords: semantic interoperability in production, intelligent factory environment, ontology, semantic web services, knowledge acquisition.

*S.F. Pileggi and C. Fernandez-Llatas (Eds.), Semantic Interoperability: Issues,
Solutions, and Challenges, 71–103.*

4.1 Introduction

The Internet of Things [13] describes the ubiquitous networking of intelligent everyday objects, which communicate autonomously, exchange information and provide services. The extension of this concept specific to the production domain is referred to as the Factory of Things (FoT) [48]. The FoT includes the extended usage of ubiquitous information and communication technologies in order to reach the networking of entities in the production domain. Based on this networking, all types of information in a production environment can be collected. The resulting information explosion, however, can only be mastered using a context- and knowledge-based provision and processing of data. Furthermore, the different mechatronic capabilities of a production plant, which are encapsulated and represented as services, need to be orchestrated in order to define complex production processes.

Semantic technologies allow the formal description of data and support the semantic processing of data by machines, i.e. the interpretation of electronically stored pieces of information with regard to their content and meaning. In this way, these technologies facilitate the structured exchange of information among heterogeneous systems, resulting in a semantic interoperability.

In the context of the FoT, semantic technologies are essential to ensure a knowledge-based interpretation of information and to facilitate an efficient discovery and a dynamic orchestration of services. This affects the representation of knowledge about production plants (including its structure, the contained components and field devices, etc.), but also the semantic description of services (e.g. the underlying mechatronic functionalities, abstract message types, pre and post conditions). In this context, the issue of achieving a semantic interoperability (e.g. between the services provided by different field devices) depicts a major research challenge. However, in our opinion, a controlled vocabulary is not realistic in the production domain because there are too many different vendors adhering to their own company-specific vocabulary. This is the reason why we believe that the creation of common semantic models that represent similarity and equivalence relations between production-specific concepts and terms is needed. But especially in the very complex and almost untouched domain of industrial production, the key issues concerning the application of semantic technologies lie in the acquisition of knowledge and the updating of the semantic models in an efficient manner. The variety of heterogeneous knowledge sources, often produced in different steps of the production plant life cycle, makes these issues even more difficult.

To address these problems, we describe our conceptional approach to model production plants using semantic technologies on the basis of existing domain-specific knowledge sources and knowledge acquisition techniques. To this end, existing procedure models, which describe the different steps that are essential to build up semantic models, are evaluated and adapted. We believe that the following steps are essential: identification and inspection of knowledge sources, (semi-)automatic processing of these knowledge sources, definition of the semantic plant model based on methods for automatic knowledge acquisition, extension of the model with lexical knowledge (e.g. similarity relations between technical terms and abbreviations) mapping different production-specific terminologies. The development of semantic models based on domain-specific knowledge sources in combination with lexical knowledge depicts a first step towards semantic interoperability in industrial production, which is essential to facilitate an automated information exchange among heterogeneous systems. We tested and evaluated our methodology in a case study, in which the development of a semantic plant model serves as the basis for the dynamic discovery and orchestration of semantically described services within an industry-related, intelligent production environment, the *SmartFactory*[KL] [47] in Kaiserslautern.

This chapter extensively describes the scientific and technical background about the Internet of Things, the Factory of Things, service-oriented architectures, semantic technologies, semantic web services and knowledge acquisition methods in the following section. After that, the most important related work concerning ontology development methods, the usage of knowledge acquisition methods in the production domain and the application of semantic web services are discussed. As the main topic of this chapter, we introduce our concept for the semantic modeling of production plants based on existing knowledge sources, lexical knowledge and methods for (semi-)automatic knowledge acquisition. This concept is meant as a first step towards a general methodology to create semantic models that help to realize a semantic interoperability in industrial production. The subsequent section illustrates the hands-on application of our conceptional approach to a case study performed within a real-life production plant. The developed semantic model serves as the basis for an efficient discovery and dynamic orchestration of services to build production processes. Finally, we present our conclusions and discuss opportunities for future work.

4.2 Background

This section describes the basic ideas of the Internet of Things and the Factory of Things in order to make the research field in which the approach described in this chapter is resided more clear. In addition the most important information about service-oriented architectures, semantic technologies, semantic web services and knowledge acquisition methods are presented, which is meant to give the reader of this chapter the necessary background knowledge to understand the presented methodology and case study.

4.2.1 The Internet of Things

The Internet of Things [13] describes the ubiquitous networking of intelligent everyday objects, which communicate autonomously, exchange information and provide services. The technological basis for making the idea of such a ubiquitous network of auto-organized intelligent entities reality is formed by the increasing capabilities of microelectronics, new communication protocols like IPv6 and Auto-ID technologies (e.g. RFID, NFC). In addition, sensor technology integrated into the smart objects makes it possible to capture relevant information from the environment and transfer it to other entities in the network. Following the vision of the Internet of Things, a seamless linking of the real, physical world with the digital world of data will be realized. While some technologies of the Internet of Things have already started to find their way into our everyday life (e.g. in logistics, smart homes, smart cars), the adaption of new information and communication technologies in industrial production is still in the early stages of development.

4.2.2 The Factory of Things

The extension of the concepts of the Internet of Things specific to the production domain is referred to as the Factory of Things [48]. This vision of a future factory focuses on the intelligent interaction of smart objects (e.g. products, plant components, handheld devices) based on semantic services to fulfil certain production-specific tasks. From a technological point of view, the major challenge to make this vision come true is to allow today's plant components to host a web server, through which their mechatronic functionality can be provided as a web service. As shown in Section 4.5.4, this can already be done today. However, in order to build up whole production plants composed of intelligent field devices that collaboratively control the production process, several research challenges have to be tackled. These challenges

are concerned with the definition of the underlying service-oriented architecture, especially the identification of suited levels of service aggregation,
the semantic description of services and the efficient discovery and dynamic context-based orchestration of these services. This chapter describes
a concept for the creation of semantic plant models, which depict important
building blocks of the Factory of Things.

4.2.3 Basics of Service-Oriented Architectures

Service-oriented Architecture (SOA) describes a collection of design principles for the development of distributed systems. SOA relies on the paradigm
of service-orientation, i.e. the different software components provide their
functionalities as loosely-coupled services over a network. This means that
services provide functionalities over standardized interfaces independent
from the underlying implementation. Web service technologies have emerged
as the most prevailing implementation of SOA. XML-based languages like
WSDL (Web Service Description Language) are used to describe a service's
interface, functionalities and characteristics. The most important activities in
a SOA-based system are discovery and orchestration. The UDDI (Universal
Description Discovery and Integration) specification depicts a way to build
service repositories, in which services can be published by service providers
and searched and discovered by service requesters. The combination of existing web services to more complex services or processes is usually performed
using BPEL (Business Process Execution Language).

There are several issues concerning the implementation and usage of systems based on the SOA paradigm [3]. The description of a service's interfaces
and functionalities via standards like WSDL is of a mere syntactic nature.
This means that the meaning of a service's capabilities, for instance, must
be interpreted by the service user. Furthermore, service repositories based
on UDDI do only provide a keyword-based search rather than a capability-
based discovery of services. These difficulties also affect the orchestration
of services. As service orchestration using languages like BPEL is based on
the syntactic description of services, process designers must bind appropriate
services at design time, resulting in a static orchestration.

Semantic technologies can be used to describe the meaning of service
capabilities in a machine-understandable manner by adding a semantic layer
to the syntactic service descriptions. In this way, an efficient discovery and
dynamic orchestration of services can be realized. In the following sec-

tions, semantic technologies and the different approaches for the definition of semantic web services are illustrated.

4.2.4 Basics of Semantic Technologies

Semantic technologies allow the formal description of data and support the semantic processing of data by machines, i.e. the interpretation of electronically stored pieces of information with regard to their content and meaning. The formal, explicit representation of knowledge forms the cornerstone of the Semantic Web [2] and includes both the modeling of knowledge and the definition of formal logics, which provide rules to draw inferences over the modeled knowledge base.

While several semantic modeling approaches for the representation of knowledge with different expressional power exist (e.g. taxonomies, thesauri, topic maps [12]), ontologies depict the most popular and the most powerful approach of explicit knowledge representation. Ontologies, often defined as "an explicit specification of a conceptualization" [14], enable the modeling of information as an independent knowledge base. Ontologies consist of three basic structures, namely classes (or concepts), relations and instances. In addition, restrictions, rules and axioms can be defined in order to model complex coherences.

Ontologies facilitate the structured exchange of information among heterogeneous systems, resulting in a semantic interoperability. To this end, special description languages are needed. The Web Ontology Language (OWL) is the ontology description language of the Semantic Web initiative standardized by the W3C [42]. OWL provides formal semantics and an RDF/XML-based syntax. The Resource Description Framework (RDF) is a formal description language for meta data, which describes information by means of so called RDF triples (subject – predicate – object) [25]. RDFS (RDF Schema) extends RDF by properties, domain and range constructs and hierarchical structures. It is already possible to model simple taxonomies using RDFS. However, OWL is the most expressive of these description languages. It provides additional features to describe complex logical expressions, cardinalities and axioms (e.g. disjunction), etc.

4.2.5 Semantic Web Services

As mentioned in Section 4.2.3, the description of a services interfaces and functionalities via standards like WSDL is of a mere syntactic nature. Similarly, defining an orchestration of services using languages like BPEL happens in a syntactic manner. As a result, there is a semantic gap between the syntactic description of web services and the underlying meaning. On the basis of a semantic annotation using semantic technologies, the meaning of a web service definition can be described in a machine-understandable manner. This additional semantic layer helps to enable the efficient discovery and the dynamic orchestration of services to build higher-value services or processes. Several approaches exist to describe web services, their operations and parameter types semantically [3]. The most common semantic web service technologies include SAWSDL, WSMO and OWL-S.

SAWSDL (Semantic Annotations for WSDL), the successor of WSDL-S [29], is a W3C recommendation, which describes a lightweight mechanism to add semantics to web services [19]. It is aimed at annotating certain parts of the web service's WSDL descriptions by adding references to semantic models. It does not specify any description language for the referenced semantic model, i.e. arbitrary models can be used such as UML or OWL ontologies. SAWSDL uses the *modelReference* attribute to annotate XML Schema type definitions, element declarations, portTypes, operations, and messages. In addition, the attributes *liftingSchemaMapping* and *loweringSchemaMapping* are used for the mapping between the technical representation of parameter types using XML and the corresponding semantic concepts. In contrast to other approaches such as OWL-S, there is no support for the description of preconditions and effects of a service.

WSMO (Web Service Modeling Ontology) provides a conceptional model for the description of different aspects of a semantic web service [39]. Its corresponding description language WSML (Web Service Modeling Language) is used to model WSMO services in a formal manner. WSMO defines four top level concepts, namely ontologies, goals, services and mediators. Ontologies provide the terminology that is used by the other WSMO elements. Goals describe the task a web service is required to solve. Services provide a semantic description of the web service's functional and behavioral aspects. Mediators are used to resolve mismatches between different WSMO elements.

OWL-S is a technology to describe the semantics of a web service based on OWL ontologies [26]. To achieve this, it consists of three types of knowledge about a web service: what does the service do (ServiceProfile), how does the service work (ServiceModel) and how can the service be invoked (ServiceGrounding). As opposed to SAWSDL, which follows a bottom-up approach to annotate services, OWL-S models the semantic service description on a high level of abstraction as an OWL ontology and then links this description to the concrete WSDL file. This approach is very expressive and makes it possible to describe highly powerful semantics of a service. However, at the same time, this increases the complexity of the description process significantly.

OWL-S not only offers ways to describe single web services semantically, but also to model processes composed of single web services. To this end, OWL-S distinguishes between so called Atomic Processes, Composite Processes and Simple Processes. An Atomic Process can be invoked directly as it is connected to a concrete WSDL file. It corresponds to the operation of a WSDL service. A Composite Process can be composed of Atomic Processes or further Composite Processes. The composition is described by OWL-S control constructs like Sequence, Split, and so on. A Simple Process represents an abstract template of a service, i.e. it depicts a semantic description of a service without concrete linking to a WSDL file. However, it can be assigned to an Atomic Process at runtime using the *realizedBy* reference.

4.2.6 Knowledge Acquisition Methods

The issue of knowledge acquisition is commonly known to be one of the major constraints in the development of knowledge based systems. Moreover, the great challenge lies in identifying appropriate knowledge sources and in developing suitable methods to extract the contained knowledge. Knowledge acquisition involves the collection, interpretation and formalization of knowledge. In general, three possibilities can be distinguished: direct, indirect and automatic knowledge acquisition. Direct knowledge acquisition means that the domain expert performs the formalization of the domain knowledge using modeling tools for example. In case of indirect knowledge acquisition, a knowledge engineer uses interview and observation techniques to retrieve the knowledge of the domain expert. Both direct and indirect knowledge acquisition commonly are time-consuming and costly. Furthermore, the updating of corresponding knowledge bases has to be done by these experts as well. That is why these approaches are not suitable for the creation and updating of semantic models in industrial production.

Several approaches exist to break the "knowledge acquisition bottleneck" [31] using methods from information retrieval, information extraction or ontology learning for instance. Approaches based on information retrieval are concerned with the search for information within documents or for documents within large collections (e.g. databases or the Web) [41]. The problem with information retrieval methods is that there is a large amount of information, some of which is irrelevant because the meaning of the information is not understood by the machine. Information extraction facilitates the extraction of structured information from textual collections using natural language processing methods in most of the cases [6]. A more advanced approach is the concept of ontology learning [4]. The goal of ontology learning is the extraction of terms, relations between these terms and even axioms from plain text to support the (semi-)automatic creation of ontologies. Ontology learning builds upon technologies for information retrieval and information extraction and involves additionally natural language processing and machine learning techniques. Wong et al. [45] give an extensive overview about different methods and tools for ontology learning.

4.3 Related Work

This section discusses the relevant related work concerning ontology development methods, production-specific knowledge acquisition approaches and the usage of semantic web service technologies for discovery and orchestration purposes in the production domain.

4.3.1 Ontology Development Methods

The creation of semantic models such as ontologies is a complex and time-consuming task. That is why several approaches have been made to develop procedure models and methodologies that define the different steps and activities necessary for the efficient creation of semantic models. Pinto and Martins [36] as well as Corcho et al. [5] give extensive overviews of the different ontology development approaches. The most important methodologies are briefly summarized in the following.

Uschold and King developed the Enterprise methodology [44], which includes four main activities, namely identification of the purpose of the ontology, building the ontology, evaluation and documentation. According to the authors, the building activity should cover the capturing of knowledge, the coding in a formal language, and the reusing of knowledge by integrating existing ontologies.

The Tove methodology proposed by Grüninger and Fox [15] focuses on the creation of first-order logic models. In a first step, the scope of the ontology is determined by defining problem scenarios and intended solutions. After that, so called competency questions are used to identify queries that the ontology should be able to answer. Based on these competency questions, the terminology of the ontology is specified, i.e. necessary concepts and relations are identified. Then, the competency questions are formalized in first-order logic expressions. Before evaluating the competency and completeness of the ontology, formal axioms, i.e. necessary and sufficient conditions for the terminology, are described.

Methontology [9] is a general procedure model for the creation of ontologies, which is influenced by software engineering methods. It identifies the most important tasks of ontology building: specification, conceptualization, integration, formalization, implementation and maintenance. The activities of knowledge acquisition, evaluation and documentation should be performed during the whole development process. The different tasks are specified regarding techniques and tools that should be used. This methodology is agreed to be the most mature by many experts (e.g. [5]).

Even though these methodologies depict cornerstones in the development of ontologies, they only consider partial aspects of the building process in detail. In particular, the topic of knowledge acquisition is not discussed sufficiently. In most of the cases, only a purely manual knowledge acquisition (e.g. expert interviews) is considered. We believe that the definition of a formally defined development process based on existing procedure models, but customized for the needs of the very special domain of industrial production is still an important research issue. Such a process particularly must take the topics of knowledge acquisition and knowledge update into account because these are the major constraints in the development of knowledge based systems. Efficient solutions to these issues are essential to convince industrial adopters.

4.3.2 Production-Specific Knowledge Acquisition Approaches

There have been only a few works dealing with automatic knowledge acquisition in the production domain such as [1, 32] or [46]. However, these approaches either focus on very special types of knowledge sources or do not employ technologies that are powerful enough to extract semantic relationships between concepts. The most interesting approach is the methodology by Li et al. [22] aimed at acquiring engineering knowledge from several heterogeneous knowledge sources. The developed ontology serves as the basis for an information retrieval framework, which helps to find engineering content for product design in an effective manner. The authors convert engineering documents such as technical reports or CAD drawings into text files in a first step. The resulting unstructured knowledge is then converted manually into semi-structured templates (so called "knowledge worksheets"), which can be parsed into the ontology. Thus, a distinction is made between concept acquisition, relationship acquisition and lexicon acquisition. The lexical terms are obtained by querying WordNet [28]. After the basic structure of the ontology is generated, the rest of the ontology is modeled using the ontology editor Protégé [33]. In a final step, the completeness and accuracy of the ontology is validated in several experiments. To the best of our knowledge, this is the most promising approach of semi-automatic knowledge acquisition in the domain of production. Although there is still a lot of manual effort necessary in their creation process (around 100 hours for acquisition and formalization of the knowledge), the underlying concepts form an important basis of our approach. Nevertheless, Li et al. focus on improving information retrieval in product design, while we emphasize the creation of a semantic plant model to reach semantic interoperability. Furthermore, we focus on an improved technique to add similarity relations to the ontology and on approaches towards automatic updates of the semantic model.

4.3.3 Semantic Web Services in Industrial Production

While in several fields of application a central role is assigned to semantic technologies, their usage has not gained acceptance in the field of industrial production yet. The main reasons for that might be the lack of illustrating application examples as well as the complexity of the technical implementation in production environments. Nevertheless, within the last years more and more research has been carried out on the application of semantic technologies to the production domain. Ontologies are frequently used to build a common interaction model for agents operating the production process [8] or for a flexible reconfiguration of assembly systems [23]. With the evolving usage of service oriented architectures in production systems, the semantic description of services becomes more important. Particularly, within the scope of the SOCRADES project, several concepts for the usage of ontologies in the production domain [27] and for the semantic discovery and orchestration of services to production processes [21] have been developed. These concepts are closely related to our approach. However, they only show the feasibility of semantic service implementations in small setups or simulations. We emphasize on the implementation of industry-related experimental setups within a real-world research facility, the *SmartFactory*KL. In addition, we believe that clearly defined methodologies for the creation of semantics-supported systems in factory automation are necessary in order to make semantic technologies, especially semantic services, applicable to the production domain.

Concerning the dynamic discovery of services, there have been several approaches to extend and improve commonly used syntactic UDDI repositories (e.g. [7]). In particular Kourtesis et al. [20] present a service description and discovery with SAWSDL. However, these works neither take the specifics of the production domain into account nor do they implement their findings in production-related experimental setups. Even more importantly, the focus of these works is only on discovery rather than on an appropriate level of semantics to form the basis for a dynamic orchestration of services. Guinard et al. [16] describe a concept for the discovery and selection of real-world services provided by embedded devices. Similar to our approach, they build upon the results of the SOCRADES project concerning the DPWS networking discovery mechanism. In addition, they discuss a concept for on-demand provisioning of missing services, which depicts an interesting idea for a future extension of our system. Samaras et al. [40] describe a concept to integrate resource constrained devices into enterprise information systems. To this end, they propose a discovery mechanism based on semantic annotations. The central topic of their work is the adaption of existing web service standards to work with resource restricted devices. Neither work focuses on the usage of semantic services within the domain of factory automation as the basis for a dynamic orchestration.

There have been several works dealing with the orchestration of services in factory automation. Many of them (e.g. [18, 37]) propose a purely syntactical orchestration of services based on BPEL, for instance. Indeed, these approaches are better suited for static workflows than for a dynamic orchestration of services [38]. In particular, Ferrándiz-Colmeiro et al. [10] propose a concept closely-related to our vision of a dynamic orchestration of services to adapt production processes. They describe an orchestration system which uses ontologies to transform abstract process models to concrete BPEL processes tailored to the capabilities of the respective industrial plant, which are represented as services in a UDDI repository. This work depicts an important basis for the future research on semantic orchestration of services in the production domain.

4.4 Towards Semantic Interoperability in Production

In this section, we present our ideas towards semantic interoperability in production. After having described the interoperability issues in the production domain in more detail, we are going to illustrate our conceptional approach for the creation of semantic plant models based on domain-specific knowledge sources and lexical knowledge extraction.

4.4.1 Interoperability Issues in Industrial Production

As defined by the IEEE glossary, interoperability means "the ability of two or more systems or components to exchange information and to use the information that has been exchanged" [34]. In order to facilitate an optimal usage of the exchanged information, these systems have to understand the meaning of information. Whereas existing syntactic approaches (e.g. using web services) are sufficient to realize a technical interoperability, a semantic interoperability can only be achieved by using sophisticated semantic models.

Because of the growing usage of modern information and communication technologies, the amount of available information increases continuously in the domain of industrial production. This causes the complexity of information management and distribution to increase too. At the same time, information sources are represented in a heterogeneous, distributed manner. These information sources are often created by several technical crews during different phases of the production plant life cycle (e.g. planning, operation, maintenance). As a result, equal or similar information is described in a different way within these information sources, which can lead to inconsistencies and redundancies. A formal description of the meaning of information is missing, which is why misinterpretation by the involved human actors or systems can occur. Consequently, interoperability among systems in production environments, but also among software tools is a key issue in industrial production.

Following the vision of a Factory of Things [48], the usage of ubiquitous information and communication technologies in order to reach the networking of entities will further increase. To make such a ubiquitous networking of components, field devices, applications, services and humans possible, it is essential to ensure a semantic interoperability among these entities. We believe that a semantic plant model, which integrates the complex technical information from different knowledge sources, depicts a promising approach to act as a semantic interoperability model in industrial production. In the

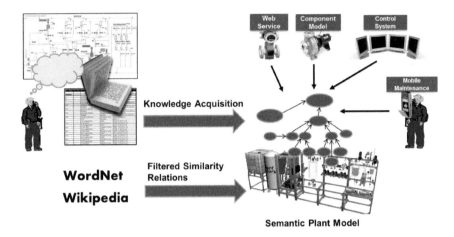

Figure 4.1 Overview of the approach to create semantic plant models.

next section, we present a conceptional approach for the creation of such a semantic plant model.

4.4.2 Conceptional Approach for the Creation of Semantic Plant Models

The development of semantic models based on domain-specific knowledge sources in combination with lexical knowledge depicts a first step towards semantic interoperability in industrial production, which is essential to facilitate an automated information exchange among heterogeneous systems.

As described in the last section, interoperability depicts a key issue in industrial production. In this section, we present a conceptional approach (as shown in Figure 4.1) for the creation of a semantic plant model, which integrates knowledge from several heterogeneous sources about components of a production plant, their coherences, functionalities and similarity relations, for example. Such a semantic plant model can ensure semantic interoperability between heterogeneous systems by modeling equivalence and similarity relations between technical terms and abbreviations. As a result, the different systems and models in an intelligent production environment such as semantic services running on field devices, component models, control systems or mobile devices used by maintenance workers can communicate on a semantic level.

The major challenges that must be addressed when building such a semantic plant model are the following: selection of a procedure model to define a basic development process, identification of appropriate knowledge sources, adaption of suitable methods for knowledge acquisition to support the modeling process, extension of the model with appropriate equivalence and similarity relations, and maintenance and update of the model. Our ideas and concepts to address these challenges are described in the following sections.

4.4.2.1 The Procedure Model

As described in the last section, several issues must be addressed in order to make the creation of semantic models applicable in industrial production. We believe that a clearly defined formal methodology is needed, which especially provides solutions to the problems of knowledge acquisition and knowledge update. The basic structure of such a methodology can be built by existing procedure models as described in Section 4.3.1. We investigated the different existing approaches and found that the METHONTOLOGY life-cycle [9] is the most appropriate to form the basis for the creation of the semantic plant model. The main reason for choosing METHONTOLOGY is its maturity [5] and its extensive description of the knowledge acquisition activity spanning over the whole ontology development life-cycle. However, in our approach, we focus less on the acquisition of knowledge via expert interviews, but more on the extraction of concepts and relations from production-specific knowledge sources. As depicted in Figure 4.2, we believe that the activities of knowledge acquisition and maintenance are the most critical in the development of semantic models in the production domain. Thus we are going to focus on ideas, concepts and implementation to improve these activities in our research.

4.4.2.2 Analysis and Selection of Knowledge Sources

The issue of knowledge acquisition is commonly known to be one of the major constraints in the development of knowledge based systems. In the domain of industrial production, this issue becomes even more difficult because of the large variety of heterogeneous knowledge sources (e.g. industrial standards, manuals, specifications, diagrams, stocklists or CAD models), their semantic complexity and varying syntax (e.g. domain-specific technical terms, abbreviations). The fact that these knowledge sources are often produced by different technical crews in different steps of the production plant life cycle, which can

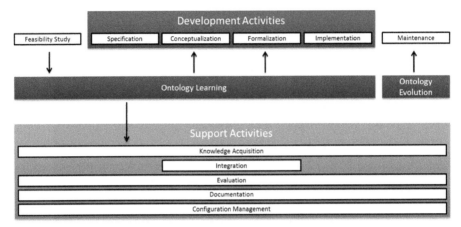

Figure 4.2 Procedure model for the creation of semantic plant models based on Methontology.

lead to inconsistent representation of the same piece of information, makes the task of semantic interoperability even more complex.

Within the *SmartFactory*[KL], an industry-related demonstration facility, we have several knowledge sources about the production plant available: industrial standards, manuals of the different field devices, circuit diagrams, piping and instrumentation diagrams, 3D and CAD models. We analyzed these knowledge sources in-depth in order to identify the potential knowledge that can be used to build up a semantic plant model. The results showed that industrial standards are of value to retrieve standardized technical terms and appreviations. The manuals of the field devices vary in their quality depending on the respective company. Indeed, most of the manuals give useful information about the components of a field device, possible error cases and the corresponding steps to fix these errors. This information could be valuable especially when developing maintenance assistance systems, which use the semantic plant model as a knowledge base. Both the circuit and the piping and instrumentation diagrams describe information about the structure of the plant and the relations (physical and process-related) between the different modules and components of the plant. 3D and CAD models can be interesting to determine the parts a plant component is composed of. However, it depends on the quality of these models. For example, information contained in CAD models can only be useful for building a semantic plant model if the CAD parts are annotated appropriately or if standard parts are used.

Following the process by Li et al. as described in Section 4.3.2, we extracted the concepts and relations needed for the semantic plant model from these

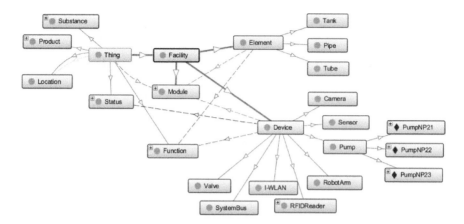

Figure 4.3 Basic structure of the semantic plant model.

knowledge sources. However, the time needed for investigation, analysis and processing of the knowledge sources shows that more advanced techniques to automatize the knowledge acquisition process are needed. As stated by Maedche and Staab [24], sophisticated techniques already exist in the fields of information extraction, machine learning or natural language processing. Indeed, the key issue is to identify and adapt adequate methods for the different complex knowledge sources in the production domain and to capture not only the domain-specific terminology, but also the relationships between technical terms as well as their meaning in a certain context. In particular, we are going to investigate to which extent existing techniques and tools for ontology learning can be applied to the production-specific knowledge sources.

4.4.2.3 Modeling of the Semantic Plant Model

We built a basic structure of the semantic plant model as shown in Figure 4.3 using Protégé 4.1. Basically, a facility is hierarchically structured into production modules, elements (e.g. tanks, pipes) and devices (e.g. sensors, pumps). The device concept contains the different device categories, which can be instantiated with the concrete field devices that exist in the plant. Elements and devices can provide different kinds of functions (e.g. convey, detect, read). In addition, we modeled concepts for representing the status of plant components, their location in the plant, and specifications of the product to be produced.

Based on the concepts and relations retrieved from the knowledge described in the last section, we refined this basic model with respect to the concrete structure of the *SmartFactory*^{KL}, the contained field devices as well as their technical properties and coherences. Because our research is also concerned with the efficient support of maintenance processes based on the provision of filtered information based on the situation of the worker, we added knowledge about potential errors and instructions to remove these errors or repair defective field devices to the plant model.

4.4.2.4 Semi-Automatic Extraction and Rating of Similarity Relations

A semantic interoperability can only be achieved if the underlying semantic model is powerful enough to understand the semantic concepts used by the models of different systems and components. As a controlled vocabulary is not realistic in the production domain because there are too many different vendors holding their own company-specific vocabulary, the representation of similarity and equivalence relations between production-specific concepts is necessary. To this end, we developed a system for the semi-automatic extension of ontologies with similarity relations, more precisely synonyms, hypernyms and hyponyms retrieved from WordNet and Wikipedia. In our system prototype, ontologies (loaded as OWL files) are displayed as a tree structure on the left-hand side. By selecting the nodes of the tree, both the WordNet and Wikipedia databases are queried for information about the corresponding semantic concepts. In the next step, the user can choose between the different meanings found by WordNet and Wikipedia and add the selected entries as similar or equivalent concepts to the ontology. To retrieve data from the WordNet database, we used the APIs JAWS 1.3[1] and JWNL 1.4.[2] To access the content and the structure of Wikipedia, we used the WikipediaMiner Toolkit 1.1 [30]. A nice feature of this toolkit is the possibility to retrieve a score of the relatedness of concepts in Wikipedia.

Indeed, the major problem is that neither WordNet nor Wikipedia are fully qualified to act as a knowledge source for the extension of semantic plant models because they cover general lexical terms rather than special engineering terminology. Simply retrieving the similarity relations to a given concept from these sources would reveal a large number of irrelevant or even incorrect entries. Therefore, we developed a scoring method, which

---

[1] Java API for WordNet Searching (JAWS): http://lyle.smu.edu/~tspell/jaws/index.html

[2] Java WordNet Library (JWNL): http://sourceforge.net/apps/mediawiki/jwordnet/index.php

takes different industrial guidelines and standards in PDF format as input and counts the number of occurrences of a given similarity relation. Following this approach, we are able to compute a domain-specific scoring of similarity relations retrieved from WordNet and Wikipedia. This makes it possible to filter out a large number of inappropriate synonyms and to add a scoring value of the chosen similar concepts.

4.4.2.5 Automatic Update of the Semantic Plant Model

The production downtime of industrial plants adds up to 2 to 8% of the yearly operation times [11]. The resulting maintenance processes often cause considerable changes of the plant hardware. As a result, the high dynamics of the production environment make it necessary to update semantic plant models continuously. As this is almost impossible to be solved manually in industrial production, new ways of recognizing or documenting the respective changes and of updating the underlying semantic models in an automatic manner must be found. We believe that advanced ways for maintenance documentation using Semantic MediaWikis, for example, must be developed. By having documentation texts annotated with semantic concepts, it would become possible to recognize the parts of a semantic model that might need to be changed accordingly. For the automatic update of the semantic model, however, techniques from the field of ontology evolution and ontology learning have to be evaluated, which depicts an important part of our future work.

4.5 Case Study: Semantic Service Orchestration for Production Processes

The growing competition between manufacturers demands a higher versatility of factory automation technology. Principles such as component based architectures (CBA) or service-oriented architectures propose an opportunity to address these challenges by encapsulating the functionality of mechatronic components in an abstract way. The implementation of a service-oriented approach using standard web technologies depict a promising possibility in production automation [35]. Service discovery and orchestration are the key functionalities of such systems. However, standard web service technologies alone are not suited for the creation of highly-flexible automation systems. In this section, we describe a semantic service discovery and orchestration system, which is based on different semantic service technologies, and provide a concept towards the creation of adaptive production processes.

The semantic plant model, which is developed following the approach described in Section 4.4, forms the basis for the semantic description of services and the interoperability among the different subsystems participating in our experimental setup.

4.5.1 Service-Oriented Production Automation

The semantic discovery and orchestration system described in this chapter builds upon an architectural approach characterized by the concept of service-oriented architectures in automation (SOA-AT) [35]. The idea of SOA-AT is to achieve a holistic communication architecture in factory automation by grounding on the general concept of service-oriented architectures. For applying the SOA concept for industrial automation systems all control functions within the automation network have to be encapsulated as services. The two basic aspects of a SOA-AT automation system are the unification of communication interfaces and the functional encapsulation of mechatronic and control functions to services. The basic functions of the manufacturing equipment that execute and monitor the production process are represented as basic services. These basic services can be aggregated to composed services for implementing control programs. In order to enable a dynamic discovery and orchestration of services, the interfaces and capabilities of these services have to be described using semantic technologies.

4.5.2 Semantic Discovery of Services

The dynamic discovery of services depicts a major challenge to enable the efficient reuse of services and the (semi-)automatic adaptation of production processes. This section describes our approach to facilitate such a dynamic discovery of services in factory automation based on semantic service descriptions.

4.5.2.1 Conceptional Approach

The semantic description of services provides the basis of a dynamic service discovery because of the following advantages: First, the meaning of a service definition can be described in a machine-understandable manner. Second, the relationships between production components and the services they provide can be described explicitly. Therefore, it is possible to find services with equivalent or similar capabilities. Third, the possibility to perform a logical reasoning over semantic models like ontologies allows the automatic interpretation of the coherences between service descriptions and the derivation of implicit knowledge.

As described in Section 4.2.5, several technologies to describe services semantically exist. We decided to use OWL-S for the description of the semantics of our services because of its expressive power (e.g. modeling of preconditions and effects), its possibility to model semantic processes, and its easy integration with OWL ontologies. For an efficient discovery and reuse of services in factory automation, there are three types of necessary information: information about the device itself (device category, device type, manufacturer, serial number), information about its provided services (service name, semantics of its capabilities, name, types and semantics of the message parameters), contextual information (physical location of the device, relationships with other devices). Information about the device and its provided services should be directly connected to the device in order to make an automatic registration of a device's services in the network possible. How this information is encapsulated within the semantic service description is illustrated in Section 4.5.2.2. Both the semantic representation of the production plant and contextual information are modeled in an ontology. Additionally, this ontology acts as a service repository to which the services can register automatically. The basic structure of the ontology and the corresponding creation process are explained in Section 4.5.2.3. For the automatic registration of devices in the network, we make use of the discovery mechanism of DPWS (Device Profile for Web Services). The network discovery of available devices provides the basis for the dynamic discovery of semantic services. The discovery process and the overall architecture of our discovery system are described in Section 4.5.2.4.

4.5.2.2 Semantic Service Description with OWL-S

For the basic discovery of a service, we focused on the semantic annotation of the service's capabilities and its device category. In addition, the message parameters of a service must be described in a semantic manner. This semantic description specifies the flow of data between inputs and outputs of different services and allows an automatic matching of the respective parameters. Take, for example, a camera service, which we implemented within the scope of the experimental setup described in Section 4.5.4. The service is provided by an industrial camera, which is able to count pills put into a container via image recognition. In this case, the service takes a *Color* as input and outputs the respective *NumberOfPills*.

For each service, an OWL-S description is created using Protégé 3.2.1 and the OWL-S Editor plugin. Thereby, the WSDL2OWL-S Tool is used initially to create the basic structure and the grounding of the OWL-S service. The services' abilities and device categories are annotated with semantic concepts modeled in our plant ontology described in Section 4.5.2.3. The input and output parameters are also linked to the respective ontological concepts using the *parameterType* reference.

In addition, ways to uniquely identify a device are needed. We decided to use a combination of manufacturer, device type and serial number as a unique identifier for devices in a manufacturing plant. With the help of such semantic service descriptions a device can automatically register its services to the semantic service repository.

4.5.2.3 Creation Process and Structure of the Ontology

The OWL ontology is divided into three modules: the service repository, the domain model, and the plant model. The service repository principally consists of the *Service* concept, which collects all the available services as ontological instances, which are automatically defined using the information extracted from the services' OWL-S files. These services have different capabilities available represented as instances of the class *Operation* (e.g. *CountPills*). Additional information such as the service name or service address is stored as properties of the service instance. Figure 4.4 shows a small segment of the semantic service repository.

As described in Section 4.4.2.3, the basic semantic plant model represents the concrete structure of the plant, its components, devices and their coherences. For each device that is newly integrated into the plant and that provides a service, an instance is added automatically to the respective device concept in the ontology. In addition, each service is assigned to the concrete device providing the service, thereby connecting the service repository with the plant model. In order to also discover equivalent or similar services based on logical reasoning, an explicit representation of equivalences and similarities is needed. To this end, we used our tool described in Section 4.4.2.4 to add synonyms, hypernyms and hyponyms of the corresponding concepts or instances to the ontology in a semi-automatic manner.

The domain model contains domain specific knowledge, which can be used for a logical reasoning over the suitability of a service. For instance, the capabilities of different proximity sensors are modeled, which allow to automatically decide (depending on the material of the object that has to be detected) whether the respective sensor service is suitable for the task at hand.

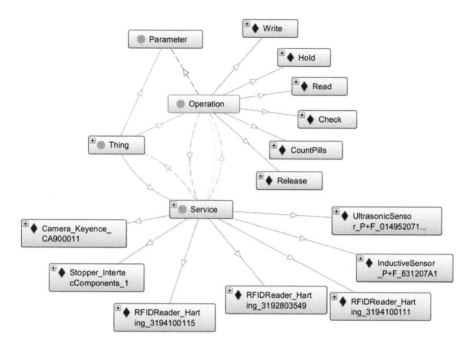

Figure 4.4 Basic structure of the semantic service repository.

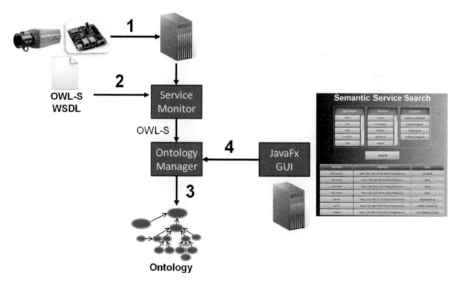

Figure 4.5 System architecture of the semantic service discovery.

4.5.2.4 Discovery System Architecture

Figure 4.5 depicts the architecture of our semantic service discovery system. The numbers represent the steps of the discovery process. Each field
device or production component stores its semantic service description file,
therefore having all the information about its device category, its services
and interfaces directly available. With the help of the discovery mechanism
of DPWS (DPWS4J API), the Service Monitor, a java program running on
a server in the network sends an initial probe message to find all devices
already on the network. In addition, a device can broadcast a hello message
when being available (step 1 in Figure 4.5). The Service Monitor retrieves the
device's DPWS metadata (e.g. location of the OWL-S file) and passes it to
the Ontology Manager. This function library, which is based on the OWL API
3.2.2 [17] and the FaCT++ reasoner [43] (version 1.5.0), is used to extract all
information that is relevant for registering the service (e.g. device category,
operations of the service, parameter types) from its OWL-S file. The corresponding service instance is then copied to the ontology (step 3 in Figure 4.5).
Using this mechanism, the ontology acts as a semantic service repository
and contains only those services that are actually available. Therefore, it can
be used as a knowledge base for the dynamic discovery of services. The
discovery process can either be performed automatically by other services
or manually by a process engineer, for instance. By providing the user with
different filter values (filter by device, by operation or by location) displayed
as adaptive lists in a JavaFx GUI, the user is able to easily find the service
he/she is looking for (step 4 in Figure 4.5). By choosing the appropriate filter
values, the user defines a semantic template, which is used to make queries
against the ontology and to retrieve the matching services.

4.5.3 Semantic Orchestration of Services

As described in Section 4.5.2, the semantic service discovery depicts an important step towards a dynamic orchestration of services. In order to bring the
approach of semantic service orchestration closer to industry, we developed a
semi-automatic process design tool, which is integrated into our experimental
setup. In addition, we explain our conceptual approach towards self-adaptive
production processes.

4.5.3.1 Semantics-supported Process Design

Based on the semantic service description with OWL-S including the annotation of input and output parameters, we developed a semi-automatic service

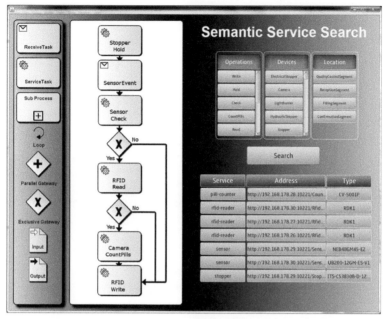

Figure 4.6 GUI of the semantics-supported process design tool.

composition prototype. This tool guides the process designer in building service orchestrations step-by-step. Figure 4.6 shows the GUI of our tool. The graphical representation of the process steps and operations is based on BPMN 2.0. On the left, the different modeling constructs are displayed, which can be placed on the canvas in the middle by drag&drop. Having placed a ServiceTask on the canvas, the editor shows the semantic service discovery frame (right-hand side of Figure 4.6). By selecting the appropriate filter values, the user can efficiently find the concrete service he/she is looking for (cf. Section 4.5.2.4). In addition to the semantic discovery, the orchestration tool gives suggestions to the designer in a proactive way. These suggestions are generated based on a semantic matching of the input and output parameters of the services integrated into the process design. For instance, the system detects whether a service needs input parameters that have not been supplied by preceding services and displays corresponding hints. When the process design is finished, the modeled BPMN process is automatically translated into an OWL-S Composite Process using the BPMN2OWL-S

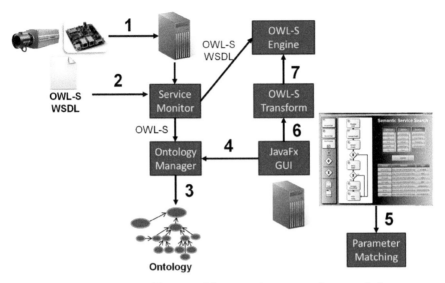

Figure 4.7 System architecture of the semantics-supported process design.

transformation algorithm that we developed and the OWL-S API 3.0.[3] This API provides an engine, which can be used to execute the composed process.

Figure 4.7 shows the architecture of the semi-automatic process design system. As the system integrates the semantic discovery system described in Section 4.5.2, the discovery part of the architecture stays the same (steps 1 to 4). The OWL-S descriptions are send additionally to the OWL-S Engine, which needs the semantic service descriptions for the invocation of the services in a later step. During the graphical design of the production process, the engineer is supported when defining the message flows between the services. To this end, the Parameter Matching component compares the semantic concepts of the parameter types (step 5). The graphical process description is then passed to the OWL-S Transform module, which transforms the modeled process into an invokable OWL-S Composite Process (step 6). In the last step (step 7), this semantic description of the process is given to the OWL-S Engine, which invokes the contained services following the modeled control constructs.

---

[3] OWL-S API 3.0: http://on.cs.unibas.ch/owls-api/

4.5.3.2 A Concept for Adaptive Production Processes

The semi-automatic orchestration described in the last section depicts an intermediate step on the way towards a dynamic adaption of production processes. Our concepts for a dynamic orchestration of services based on abstract process descriptions specific to the product to be produced are described in the following. The abstract description of the product-specific production process is modeled as an OWL-S Composite Process, which consists of abstract OWL-S Simple Processes. In addition to the message parameters, the preconditions and effects of these Simple Processes must be described semantically using OWL and SWRL, for instance, on the basis of domain-specific ontologies. Preconditions and effects represent the provided and consequent state of the context (e.g. the environment, the plant, a certain component) specific to a service execution. The semantic description of a service's preconditions allows modeling the contextual information that has to be met in order to execute the service. Having modeled the effects of services enables the automatic update of the underlying semantic plant model. In this way, we are able to integrate contextual information into the semantic description of a process, which makes it possible that the atomic services can be defined independently of the context of their execution. Such an abstract process description could be directly attached to the product (e.g. storing it on an RFID tag or a smart object). By doing so, the product carries knowledge about its production process. The concrete instance of the process is generated depending on the actual structure of the production plant and the capabilities of its components making use of our semantic discovery system. Moreover, the system can adapt the process at runtime (e.g. in case of a faulty component) by finding components that provide a similar service using the semantic discovery.

4.5.4 Experimental Setup

We developed an industry-related experimental setup as part of the demonstration facility *SmartFactory*^{KL} [47] in order to test and evaluate our concepts and technologies. The *SmartFactory*^{KL} is the first vendor-independent research and demonstration facility for the application and evaluation of smart production technologies and includes both research institutes and several partners from industry. It operates a hybrid, modular demonstration plant in a 200 square meters industrial facility. Figure 4.8 shows a part of the modular production plant.

We deployed a production module as a small-scale test bed for the semantic discovery and orchestration system developed in our case study. As

Figure 4.8 Part of the *SmartFactory*^{KL} production test bed.

shown in Figure 4.9, the setup contains different industrial equipment. All these field devices can be controlled via services using micro controllers (NGW100) as a gateway. The services are implemented using web services (WSDL) generated with the toolkit gSOAP 2.7.17 and using the DPWS stack built with DWPScore 2.3.2. The sample process realizes the filling of a container with a predefined number of pills. After filling the container, the correct amount of pills will be verified by a camera system using image recognition to count the number of pills. Also, product-related information like order information or state of progress in production is stored on a RFID tag attached to the product.

Even though the number of devices and services in this setup is quite manageable, the distinction of all services is only possible by using the address of the services. Compared to productive working plants, the demonstration facility of the *SmartFactory*^{KL} with about 200 square meters, over 200 devices, eight PLCs and several different communication technologies is still relatively small. Nevertheless, searching for devices or services on a syntactical base is already time-consuming and arduous. A semantic representation of the

Figure 4.9 Experimental setup with service controlled field devices.

services allows a superior computerized support when a reprogramming or a reconstruction is necessary. Having this demonstration module including the different hands-on tools available, we aim at bringing semantic technologies together with the approach of service-oriented architectures in automation closer to industry and therefore to a more wide-spread adoption.

4.6 Conclusions and Future Work

We believe that the usage of semantic technologies in the production domain can make a valuable contribution towards managing the growing complexity, increasing the flexibility of production processes and establishing a semantic interoperability among heterogneous systems. To this end, common semantic models that represent similarity and equivalence relations between production-specific concepts and terms are needed. However, the key issues concerning the application of semantic technologies in the domain of industrial production are knowledge acquisition and knowledge updating.

In this chapter, we presented a conceptional approach for the development of semantic plant models on the basis of existing domain-specific knowledge sources, lexical knowledge and knowledge acquisition techniques. The resulting semantic plant model serves as the basis for the efficient discovery and dynamic orchestration of semantically described services to build production processes. In addition, the semantic plant model is already used for the dynamic information provision and assistance during maintenance processes in the *SmartFactory*[KL].

Because the knowledge acquisition process is still too time-consuming, we are going to investigate more improved technologies for the processing of the different types of partially unstructured engineering documents. Indeed, the key issue is to identify and adapt adequate methods for the different complex knowledge sources in the production domain and to capture not only the domain-specific terminology, but also the relationships between technical terms as well as their meaning in a given context. In particular, we are going to evaluate technologies and tools for ontology learning. These technologies could also provide possibilities to perform a more efficient updating of semantic plant models.

Concerning the semi-automatic extraction and scoring of lexical knowledge from WordNet and Wikipedia, we are going to evaluate our system in a statistical manner. In particular, the filtering mechanism of terms based on counting their occurrences based on production-specific standards and guidelines is a very simple approach, which can certainly be improved considerably.

In summary it can be said that the creation of semantic plant models as a basis for semantic interoperability can already be performed in industrial production in quite an efficient manner. To make this conceptional approach applicable for industrial adopters, however, a clearly defined methodology including additional easy-to-use tools is needed.

References

[1] Saeema Ahmed, Sanghee Kim, and Ken M. Wallace. A methodology for creating ontologies for engineering design. *Journal of Computing and Information Science in Engineering*, 7(2):132–140, 2007.

[2] Tim Berners-Lee. *Weaving the Web*. Harper, 1999.

[3] Sami Bhiri, Walid Gaaloul, Mohsen Rouached, and Manfred Hauswirth. Semantic web services for satisfying SOA requirements. In *Advances in Web Semantics I*, pages 374–395. Springer-Verlag, Berlin/Heidelberg, 2009.

[4] Paul Buitelaar and Philipp Cimiano (Eds.). *Ontology Learning and Population: Bridging the Gap between Text and Knowledge*, Frontiers in Artificial Intelligence and Applications, Vol. 167. IOS Press, Amsterdam, 2008.

[5] Oscar Corcho, Mariano Fernández-López, and Asunción Gómez-Pérez. Methodologies, tools and languages for building ontologies: Where is their meeting point? *Data Knowl. Eng.*, 46:41–64, July 2003.

[6] Jim Cowie and Yorick Wilks. Information extraction, 1996.

[7] Marco Crasso, Alejandro Zunino, and Marcelo Campo. Easy web service discovery: A query-by-example approach. *Sci. Comput. Program.*, 71:144–164, April 2008.

[8] Daniel Diep, Christos Alexakos, and Thomas Wagner. An ontology-based interoperability framework for distributed manufacturing control. In *Proceedings of IEEE Conference on Emerging Technologies and Factory Automation (ETFA2007)*, pages 855–862, September 2007.

[9] Mariano Fernandez-Lopez, Asuncion Gomez-Perez, and Natalia Juristo. Methontology: From ontological art towards ontological engineering. In *Proceedings of the AAAI97 Spring Symposium*, Stanford, USA, pages 33–40, March 1997.

[10] Antonio Ferrándiz-Colmeiro, Virgilio Gilart-Iglesias, and Francisco Mácia-Pérez. Semantic processes modelling independent of manufacturing infrastructures. In *Proceedings of the 2010 IEEE Conference on Emerging Technologies and Factory Automation (ETFA)*, pages 1–8, September 2010.

[11] Don Fitchett. *The True Cost of Downtime*. Business Industrial Network, 2002.

[12] Lars M. Garshol. Metadata? Thesauri? Taxonomies? Topic maps! Making sense of it all. *Journal of Information Science*, 30(4):378–391, 2004.

[13] Neil Gershenfeld, Raffi Krikorian, and Danny Cohen. The internet of things. *Scientific American*, 291(4):76–81, October 2004.

[14] Thomas R. Gruber. A translation approach to portable ontology specifications. *Knowledge Acquisition*, 5:199–220, 1993.

[15] Michael Grüninger and Mark S. Fox. Methodology for the design and evaluation of ontologies. In *Proceedings of Workshop on Basic Ontological Issues in Knowledge Sharing*, 1995.

[16] Dominique Guinard, Vlad Trifa, Stamatis Karnouskos, Patrik Spiess, and Domnic Savio. Interacting with the SOA-based internet of things: Discovery, query, selection, and on-demand provisioning of web services. *IEEE Trans. Serv. Comput.*, 3:223–235, July 2010.

[17] Matthew Horridge and Sean Bechhofer. The OWL API: A Java API for working with OWL 2 ontologies. In *OWLED*, CEUR Workshop Proceedings, Vol. 529. CEUR-WS.org, 2008.

[18] Francois Jammes, Harm Smit, Jose L. Martinez Lastra, and Ivan M. Delamer. Orchestration of service-oriented manufacturing processes. In *Proceedings of IEEE International Conference on Emerging Technologies in Factory Automation (ETFA2005)*, Cernobio, Italy, pages 617–624, 2005.

[19] Jacek Kopecký, Tomas Vitvar, Carine Bournez, and Joel Farrell. Sawsdl: Semantic annotations for WSDL and XML schema. *Internet Computing, IEEE*, 11(6):60–67, 2007.

[20] Dimitrios Kourtesis and Iraklis Paraskakis. Combining SAWSDL, OWL-DL and UDDI for semantically enhanced web service discovery. In *Proceedings of the 5th European Semantic Web Conference on the Semantic Web: Research and Applications (ESWC'08)*, pages 614–628. Springer-Verlag, Berlin/Heidelberg, 2008.

[21] Jose L. Martinez Lastra and Ivan M. Delamer. Semantic web services in factory automation: Fundamental insights and research roadmap. *IEEE Transactions on Industrial Informatics*, 2(1):1–11, February 2006.

[22] Zhanjun Li, Maria C. Yang, and Karthik Ramani. A methodology for engineering ontology acquisition and validation. *Artif. Intell. Eng. Des. Anal. Manuf.*, 23:37–51, February 2009.

[23] Niels Lohse, Hitendra Hirani, and Svetan Ratchev. Equipment ontology for modular reconfigurable assembly systems. *International Journal of Flexible Manufacturing Systems*, 17:301–314, 2005.

[24] Alexander Maedche and Steffen Staab. Ontology learning for the semantic web. *IEEE Intelligent Systems*, 16:72–79, March 2001.

[25] Frank Manola and Eric Miller. RDF primer, 2004.

[26] David Martin, Mark Burstein, Drew Mcdermott, Sheila Mcilraith, Massimo Paolucci, Katia Sycara, Deborah L. Mcguinness, Evren Sirin, and Naveen Srinivasan. Bringing semantics to web services with OWL-S. *World Wide Web*, 10:243–277, September 2007.

[27] Jose L. Martinez Lastra and Ivan M. Delamer. Ontologies for production automation. In Tharam S. Dillon, Elizabeth Chang, Robert Meersman, and Katia Sycara (Eds.), *Advances in Web Semantics I*, pages 276–289. Springer-Verlag, Berlin/Heidelberg, 2009.

[28] George Miller and Christiane Fellbaum. *WordNet: An Electronic Lexical Database*. MIT Press, Cambridge, MA, 1998.

[29] John Miller, Kunal Verma, Preeda Rajasekaran, Amit Sheth, Rohit Aggarwal, and Kaarthik Sivashanmugam. WSDL-S: Adding Semantics to WSDL, White Paper, 2004.

[30] David N. Milne and Ian H. Witten. An open-source toolkit for mining Wikipedia. In *Proceedings of New Zealand Computer Science Research Student Conference (NZCSRSC)*, Vol. 9, 2009.

[31] Martin Mozina, Matej Guid, Jana Krivec, Aleksander Sadikov, and Ivan Bratko. Fighting knowledge acquisition bottleneck with argument based machine learning. In *ECAI*, Frontiers in Artificial Intelligence and Applications, Vol. 178, pages 234–238. IOS Press, 2008.

[32] Jyotirmaya Nanda, Timothy W. Simpson, Soundar R.T. Kumara, and Steven B. Shooter. A methodology for product family ontology development using formal concept analysis and web ontology language. *Journal of Computing and Information Science in Engineering*, 6(2):103–113, 2006.

[33] Natasha F. Noy, Michael Sintek, Stefan Decker, Monica Crubezy, Ray W. Fergerson, and Mark A. Musen. Creating semantic web contents with Protege-2000. *IEEE Intelligent Systems*, 16(2):60–71, March/April 2001.

[34] Institute of Electrical and Electronics Engineers. IEEE Standard Computer Dictionary: A Compilation of IEEE Standard Computer Glossaries, 1990.

[35] Lisa Ollinger, Jochen Schlick, and Stefan Hodek. Leveraging the agility of manufacturing chains by combining process-oriented production planning and service-oriented manufacturing automation. In *Proceedings of the 18th IFAC World Congress*, 2011 (to appear).

[36] Helena Sofia Pinto and Joao P. Martins. Ontologies: How can they be built? *Knowl. Inf. Syst.*, 6:441–464, July 2004.

[37] J. Puttonen, A. Lobov, and J.L. Martinez Lastra. An application of bpel for service orchestration in an industrial environment. In *Proceedings of IEEE International Conference on Emerging Technologies and Factory Automation (ETFA2008)*, pages 530–537, September 2008.

[38] Wei Ren, Gang Chen, D. Chen, Chor Ping Low, Chengzheng Sun, Jing Bing Zhang, and Zhonghua Yang. Searching for service-oriented strategies of dynamic composition of web services: A comparative perspective. In *Proceedings of 33rd Annual Conference of the IEEE Industrial Electronics Society (IECON2007)*, pages 2615–2620, November 2007.

[39] Dumitru Roman, Uwe Keller, Holger Lausen, Jos de Bruijn, Rubn Lara, Michael Stollberg, Axel Polleres, Cristina Feier, Christoph Bussler, and Dieter Fensel. Web service modeling ontology. *Applied Ontology*, 1(1):77–106, 2005.

[40] Ioakeim K. Samaras, John V. Gialelis, George D. Hassapis, and Vincent A. Akpan. Utilizing semantic web services in factory automation towards integrating resource constrained devices into enterprise information systems. In *Proceedings of the 14th IEEE International Conference on Emerging Technologies & Factory Automation (ETFA'09)*, Piscataway, NJ, pages 610–617. IEEE Press, 2009.

[41] Amit Singhal. Modern information retrieval: A brief overview. *Bulletin of the IEEE Computer Society Technical Committee on Data Engineering*, 24(4):35–42, 2001.

[42] Michael K. Smith, Chris Welty, and Deborah L. McGuinness. OWL web ontology language guide, February 2004.

[43] D. Tsarkov and I. Horrocks. Fact++ description logic reasoner: System description. In *Proceedings of the International Joint Conference on Automated Reasoning (IJCAR2006)*, Lecture Notes in Artificial Intelligence, Vol. 4130, pages 292–297. Springer, 2006.

[44] Mike Uschold and Martin King. Towards a methodology for building ontologies. In *Proceedings of Workshop on Basic Ontological Issues in Knowledge Sharing*, held in conjunction with IJCAI-95, 1995.

[45] Wilson Wong, Wei Liu, and Mohammed Bennamoun. Ontology learning from text: A look back and into the future. *BACM Computing Surveys*, 2011 (to appear).

[46] Maria C. Yang, William H. Wood, III, and Mark R. Cutkosky. Design information retrieval: A thesauri-based approach for reuse of informal design information. *Eng. with Comput.*, 21:177–192, November 2005.

[47] Detlef Zühlke. Smartfactory: From vision to reality in factory technologies. In *Proceedings of the 17th International Federation of Automatic Control (IFAC) World Congress*, pages 82–89, 2008.

[48] Detlef Zühlke. Smartfactory – Towards a factory-of-things. *Annual Reviews in Control*, 34(1):129–138, 2010.

5

Ontology Driven Smart Space Application Development

M. Mohsin Saleemi, Natalia Díaz Rodríguez, Espen Suenson,
Johan Lilius and Iván Porres

*Department of Information Technologies, Turku Centre for Computer Science
(TUCS), Åbo Akademi University, Turku, Finland; e-mail: johan.lilius@abo.fi*

Abstract

This chapter presents an approach to create an abstraction layer and appropriate tools for rapid application development for *Smart Spaces*. The proposed framework is described together with the overall process for application development. An approach of how to integrate OWL-S grounding into the agents is described for allowing service discovery and composition. We also show results of a case study implementation to illustrate the functionality of the proposed framework. This case study shows that interoperability can be realized by agents that describe information about themselves using some common ontology.

Keywords: smart space, ontology, interoperability, application development, OWL-S.

5.1 Introduction

While the Semantic Web envisions more well-structured data enabling new possibilities for the Internet, the semantic concept is also being adopted into other areas. One of these areas is the *Smart Space*, which, although very similar, comes with a different set of restrictions, challenges and possibilities.

*S.F. Pileggi and C. Fernandez-Llatas (Eds.), Semantic Interoperability: Issues,
Solutions, and Challenges,* 105–129.

The reason is that *Smart Spaces* heavily depend on heterogeneous devices, systems and services and need to be made seamlessly interoperable to be used effectively in a device, vendor and domain independent manner. The diversity of devices and standards raises the problem of enabling interoperability of different services within different devices.

A *Smart Space* is an abstraction of space that encapsulates both the information in a physical space as well as the access to this information, such that it allows devices to join and leave the space. In this way, a *Smart Space* becomes a dynamic environment whose identity changes over time when the set of entities interact with it to share information between them. Moreover, as *Smart Spaces* provide information about a physical environment which is shared with inherently dynamic applications, ubiquitous ambient services should adapt user preferences in each particular context. These ubiquitous applications employ a range of different devices in addition to mobile phones to provide a set of innovative services that are both social and personalized. This requires more advanced methods for data handling and understating and approaches for data exchange and communication across heterogeneous sources. Furthermore, there are several other issues to be solved such as interoperability issues, common application development platforms and the development tools for rapid application development.

The main research problems we are dealing with are how to develop fast and rapid applications for *Smart Space* using the traditional Object Oriented (OO) programming approach and how to achieve interoperability in these applications. The objective is to develop a generic and comprehensive inter-operability solution that enables the devices and applications from different domains to communicate with each other and construct a scalable smart network of diverse devices.

In this chapter, we present our solutions for the given research problems. We proposed and developed user level tools, which make use of Web Ontology Language (OWL). OWL not only allows structuring the *Smart Space* content in terms of high-level programing language concept of classes but also specifies relations between the classes and their properties. Hence, entities interacting with *Smart Space* can consume and produce content according to high level OWL ontology terms. As dealing with ontologies could be very difficult for the programmers and end-users, we developed a user-level tool that generates ontology API by mapping OWL ontology concepts into Object Oriented programming language concepts. This enables application developers to create innovative *Smart Space* applications using traditional Object Oriented programming concepts without worrying about the complexity of OWL. Moreover, service discovery and composition under "unchoreographed" conditions [8] can be tackled with existing semantic web services and our proposed tools [21], which provide a solution to make easier for the programmer the combination of services differently implemented.

This chapter is organized as follows. Section 5.2 gives an overview of the related work on ontology based infrastructures. Section 5.3 gives an overview of the *Smart-M3* concept, a particular implementation of *Smart Space*. It also describes the ontology point of view for this approach and describes the tools developed for support of Knowledge Processors (KP) creation. Section 5.4 shows a case study using our framework to give proof of concepts. Section 5.5 gives a structured view of *Smart Space* and the interaction between KPs. Section 5.6 describes a service ontology and its function. It also proposes an OWL-S Python binding for service composition. Finally, Section 5.7 makes some conclusions and depicts directions for future work.

5.2 Related Work

Context-aware computing research shows a large number of context-aware systems and approaches for application development. Ontology context modeling differentiates according to simplicity, flexibility, extensibility, genericity and expressiveness [9]. Since 2004, many ontology-based systems have been developed. CoBrA and SOCAM are some examples, which use their own OWL-based approach for context processing while others like Context Managing Toolkit describe context in RDF. CoBrA [11], as agent-based infrastructure for context modeling, context reasoning and knowledge sharing, provides techniques for the user's privacy control, while Soupa and CoBrA-Ont provide some of their ontologies. SOCAM [15] introduces another architecture for building context-aware services focused on information sensing and context providers using a central server. All these systems use SQL to access the central database. In contrast, we propose RDF as a more efficient way and to restrict the queries to a smaller set of statements.

In [20], operating system concepts include context-awareness. Quaternary predicates are used for information representation, in which the fourth one is context-type. DAML + OIL is employed, as well as an MVC model. These and other projects as [28] focus basically on creating ontologies for context-representation. However, we intend to build a framework for creating context-aware development of services or applications based on the semantic architecture (which, in contrast, has a blackboard architecture).

Context Toolkit [13] is another case that enables application development through reusable components. However, its attribute-value tuples, not being meaningful enough, make application programming restricted. Another example is HIPPIE [18], which utilizes existing users' information with an awareness system to distribute context information to the users' devices. For compensating its lack of handling interaction, it was combined with NESSIE [19] which added event based awareness but still lacked semantic information description. When rules must be specified in the *Smart Space* in OWL, an OWL-Script language [25], prototyped in our group, can be considered.

When it comes to deploying services, there are several alternatives. The main ones consist of OWL-S and WSMO (Web Service Modeling Ontology). The grounding in OWL-S provides the details of how to access the service mapping from an abstract to a concrete specification of the service. WSMO approaches for grounding use mapping to XML with SAWSDL (Semantic Annotations for WSDL and XML Schema) while OWL-S utilizes WSDL (and SWRL for rules) and possibly XLST transformations. However, both groundings reduce in the end to the use of WSDL to where both OWL-S and WSMO services must be mapped for a concrete specification [7].

There are different approaches and architectures that address the issue of service composition. These approaches can be classified using several service composition features such as automatic composition [17], semi-automated composition [23], end-user interaction [24], service specification language [14], etc. In [9], the authors give a comparison of different service composition approaches. A middleware solution for end-user application composition is provided in [12]. Other approaches of flexible service composition in mobile environments are described in [10] and [26]. While existing research efforts deal with theses issues separately, there has been very limited work in ubiquitous service compositions in smart environment. In [27], the authors proposed a system consisting of a middleware and user-level tools that enable the end-users to combine, configure and control the services using their smart home devices.

Aiming at facilitating the creation of smart services we can observe that diverse technology-specific frameworks exist, but none of them results in a rapid and functional application programming tool. Comparing with previous systems, we describe an approach that tackles the challenge of context-aware ubiquitous computing using automated ontology code generation (Python and C) giving complete control over ontologies. The communication with the *Smart Space* is therefore encapsulated for the developer.

5.3 Smart-M3 Architecture

A concrete implementation of *Smart Space* is Nokia's *Smart-M3* [5], a Multi-domain, Multi-device and Multi-vendor (M3) platform consisting of a space based communication mechanism for independent agents which communicate implicitly by inserting and querying information in the space. *Smart-M3* is an open source, cross-domain architecture where the central repository of information, Semantic Information Broker (SIB) is responsible for information storage, sharing and management through the *Smart Space* Access Protocol (SSAP). SSAP provides the KPs access to the *Smart-M3* space by means of the operations: *Join/Leave* the *Smart-M3* space, *Insert/Remove* information from the SIB, *Update*, *Query* and *Subscribe* to changes.

Entities called Knowledge Processors (KPs) implement functionality and interact with the *Smart Space* by inserting/retrieving/querying common information. An application is constructed by aggregating normally several KPs where each performs a single task and communication does not happen device to device but through the *Smart Space* central repository (SIB). One device can host any number of different KPs. The application is constructed by the composition of several KPs where each KP performs a specific task. The application design in this approach differs from the traditional single device control-oriented application.

The information level interoperability provided by *Smart-M3* allows objects and devices in the physical space to define a common information representation model with Resource Description Framework (RDF). Information in the SIB is stored as RDF graphs or Triples (*Subject, Predicate, Object*).

The *Smart-M3* space is composed of one or more SIBs but even if the information may be distributed over several SIBs, the information result is the union of information stored in all SIBs associated with that space. Since SIBs are routable, devices see the same information and it does not matter to which particular SIB in a M3 space a device is connected. Figure 5.1 shows the *Smart-M3* Architecture.

This chapter presents a framework for simplifying the development of KPs agents. The *Smart Space* interface is abstracted by hiding the underlying complexity involved in ontology-driven approaches. This is achieved through a Python and C API generated from an OWL-DL ontology.

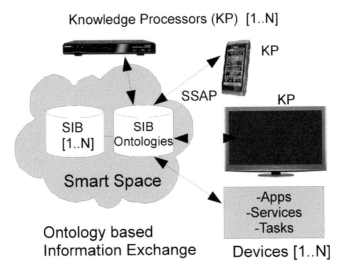

Figure 5.1   Smart-M3 architecture.

5.3.1 Ontologies in *Smart-M3*

The information interoperability needed in *Smart Spaces* can be realized by agents that describe their information using a common ontology. In the last version of the Ontology Web Language, OWL 2, ontologies can also be viewed as RDF graphs, i.e. the structural form is mapped to the RDF graph form and vice versa [2].

OWL ontologies can define the context information representing data directly obtained from context providers plus inferred information from this data using inference rules. Thus, we chose ontology based context modeling for *Smart Space* for several reasons. Firstly, the platform *Smart-M3* provides an interoperability architecture based on ontology models with support for RDF graphs storage, code generation and ontology reasoning; secondly, we chose ontology based context modeling for the fact that ontologies are the most promising and expressive models satisfying information interoperability requirements. Moreover, ontology based models provide flexibility, extendibility and genericity, key factors in context-aware ubiquitous spaces [9].

Since the main design goal for our research is the rapid and easy application development for *Smart Space* environments, the dynamic nature of OWL is something from which we can definitely benefit in the modeling.

Ontologies enable the expression of information and relations in an application. An ontology allows KPs to access and process the information related to their functionality from the *Smart-M3* space, consequently driving the KPs through the space [22].

5.3.2 Tool Support for Knowledge Processors creation

Our approach consists of two development tools modules:

1. The first part is a Python code generator that creates a static API from an OWL ontology [4] as illustrated in Figure 5.2. These mapping generate native Python classes, methods and variable declarations which can then be used by the application developer to access the data in the *Smart-M3* space as structured and specified in the OWL ontology. The generator loads an OWL ontology into a Java ontology model which provides interfaces for accessing the RDF graph. A reasoner is connected to the model to complete the inferred part of the ontology. The generator then lists all named classes in the ontology and the handler creates a counterpart OWL class in Python which is added to the code model. The class handler will list all properties and call the *ObjectProperty* and *DatatypeProperty* handlers which, in turn, translate every restriction that the property may have, e.g. `Cardinality` and `Range` restrictions.
2. The second component is the middleware layer which abstracts the communication with the *Smart-M3* as illustrated in Figure 5.3. Its main functionality is the handling of information in the central SIB with the generated API. This consists of inserting, removing and updating RDF Triples and committing changes to the *Smart Space*. It also provides functionality for synchronous and asynchronous querying. Our approach enables application developers to use the generated API to develop new applications without worrying about the *Smart-M3* interface as the generated API takes care of the connection to the *Smart Space* each time an object is created. From the *Smart-M3* point of view, the proposed framework simplifies the development of KPs by making the *Smart Space* interface more abstract and hiding the underlying complexity involved in ontology-driven approaches [16].

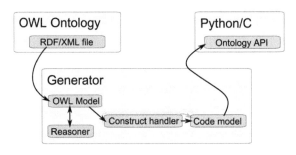

Figure 5.2 Framework overview.

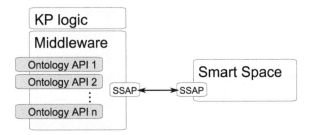

Figure 5.3 Runtime middleware for *Smart-M3*.

5.4 Case Study: From Ontology Editing with Protégé to an Application Example in Home Automation

Our demonstration scenario [16] uses a home state switch, reflecting the global state (i.e. "Home", "Away" and "Vacation"), and a heating system. Moreover, there are two additional parts for enabling interoperability: a controller and a configuration tool. In addition to these interoperating components, there is also a temperature display, and a temperature slider which can be configured to correspond to or set the different temperatures available from the heater appliance. The demonstration application consists of several KPs representing the functionality of the devices and a user interface. All devices at home connect to the SIB and insert information about themselves. A conceptual model is shown in Figure 5.4. The demonstration implementation contains a temperature service concept in addition to the house state concept shown in Figure 5.4. The temperature data service is contained in the Heater, Temperature Slider and the display. An example configuration is to set the display to show the active temperature setting in the heater.

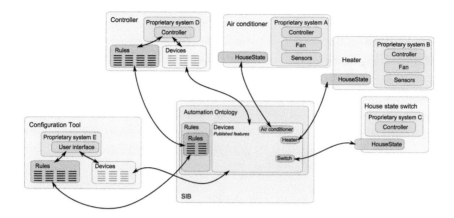

Figure 5.4 Case study overview showing an interoperability solution.

In this use-case we created an ontology containing rules for automation, concepts for expressing the house state, and the temperature. These ontology components and their attributes were edited with *Protégé* [6]. The Python Code Generator was used to generate the agent ontology API with populated instance properties.

The case study illustrates an application development approach for *Smart-M3* using our proposed framework. The KPs are developed from the generated ontology API and are able to communicate through the *Smart-M3* space providing interoperability between different devices in the example application.

All devices in the home connect to the SIB, through their respective SIB interfaces, and insert information about themselves. This information consists of a user friendly name, a list of services it provides and the data which describes its state. No automatic configuration about how they interact exist at this time. When the configuration tool is run, the user is presented with devices registered in the SIB, and can then configure rules. Rules are interpreted by a controller KP. The controller subscribes to changes in the data of the devices. In order to catch changes in state of the switch, the controller listens to new instances of the Event class. This instance contains information about what has occurred. When the controller receives a new instance of an Event, it parses through the list of rules and if there is a matching rule, it will execute the rule. In this simple implementation, a matching rule will create a new instance of the class Invoke and adds properties to it according to the configured rules. The new `Invoke` instance is subscribed to by the KP representing the service invoked, and can then be used to alter the internal state accordingly.

All devices of interest connect to the SIB through their respective SIB interfaces, insert information about themselves, the service they provide and the data which describes its state. When the configuration tool is run, the user is presented with devices registered in the SIB, and can then configure rules. Rules are interpreted by a controller KP which subscribes to changes in the data. In order to catch changes in the switch state, the controller listens to new instances of the `Event` class.

The API generator and DIEM Mediator[1] source code is available from *Smart-M3* at SourceForge [5]. The demonstration was tested with Python 2.6.x, PyQt v.4.5.4 for Python 2.6 and Nokia SIB revision 98.

The following agents in the building automation demonstration try to connect to a *Smart Space* named 'x' on 127.0.0.1 at port 10010 by default:

- `Controller.py`
- `ConfigurationTool.py`
- `HomeStateSwitch.py`
- `Heater.py`
- `TemperatureSet.py`
- `TemperatureDisplay.py`

Running python `SIB.py x` starts a SIB running locally at port 10010 with the *Smart Space* name x. The simplest use-case to run is the *HomeStateSwitch* and the *Heater*, with pre-configured addresses 17 and 7 respectively. These can be connected by the configuration tool using the following commands:

```
Command] list
Command] connect
Source address: 17
Source feature: 0
Destination address: 7
Destination feature: 0 (State might have another number)
Rule name: TestRule
```

If commands are executed correctly the heater will output its changing state following the home state switch. The KPs can be started in any order, but the

---

[1] A caching middleware for accessing the SIB

configuration tool KP does not find any devices until they are started. The suggested order is to run the controller and the configuration tool, and then any of the service providing devices or KPs. Note that subscriptions to the SIB result in a TCP timeout if no subscribed data is sent by the SIB within a quite short period of time depending on network configuration. Python cannot platform independently set TCP keep-alive messages and it is therefore recommended to run the demo locally. There are some subscriptions which are not required after running the configuration tool, thus the example might work even after this timeout error.

More information about the development tool can be found in [16].

5.5 Structured View of Smart Space

In order to explain the structure of the *Smart Space* and the role of the KPs ontologies we will consider a simple application for not missing our favorite TV program. Let us suppose that the user's favorite program is starting in few minutes according to the user profile information or fan page in Facebook and the TV guide available on the broadcaster's web page. Then the GPS in his mobile phone or his personal calendar could find out that he is not at home and start the PVR (Personal Video Recorder) to record the program. In order to address this kind of cross-domain scenario where technical and conceptual problems arise, the concept of *Smart Space* appears to encapsulate and abstract information from different services with the aim of allowing heterogeneous service composition.

A PVR could be considered as a form of API with different functions. One or several KPs can be perceived as a service, for example several KPs handling calendar activities in an application could shape a calendar service. Thus, each service acts as service provider exposing its functionality to other KPs and services through the *Smart Space*. At the same time each service acts as requester too.

In this way, we could have the PVR's KP and the mobile phone's KP connected to the SIB. Figure 5.5 shows the registered devices' KPs with their information described in their respective ontologies.

Each of these subservices within a device inform their inputs and outputs among other parameters in each of their profiles. In order to deploy the scenario of recording the favorite program, the composition of required services must be deployed in the SIB, which knows about the devices connected to the *Smart Space*. Here we find the problem that the SIB offers a persistent data repository but is a plain database giving just access to the data; no control

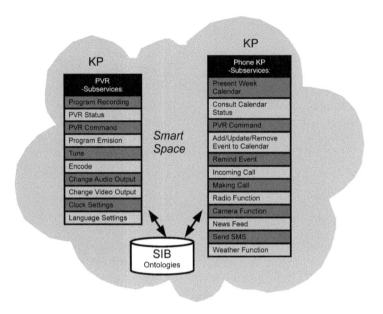

Figure 5.5 KP's services structure.

structure or computation is provided. However, if we add to the SIB a description of each subservice, all devices would be represented with a unique standard allowing language and device independent service composition. For this purpose we suggest service description with OWL-S [3] representation. The reasons are that OWL-S enables declarative advertisement of service properties and capabilities that can be used for automatic service discovery and because it describes the services in terms of capabilities based on OWL (also supported by *Smart-M3*). In addition to provide specification of prerequisites of individual services, OWL-S language describes services composition including data flow interactions [22].

By storing OWL-S instances (associated to each KP) in the SIB, the SIB would gain processing control. In this case an OWL-S interpreter running in parallel with the SIB would control the matching of compatible services when requested. Consequently, the original *Smart-M3* framework would be enhanced.

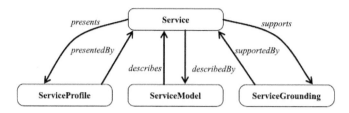

Figure 5.6 Main classes in OWL-S service ontology.

5.6 Service Ontology Description

The new generation of markup languages such as OWL was developed to support the description of specific web sites and reasoning applications. The idea behind them is to access web resources by content rather than just by keywords. When we talk about services, we will refer to local or web applications that allow actions or changes to happen in a semantic environment that enables users to locate, query, select, invoke, compose, reuse and monitor web-based services automatically [3].

OWL-S (formerly DAML-S) is formally defined as an ontology of services that allows users and software agents to interact. The OWL-S ontology is structured in three parts (Figure 5.6):

- *Service Profile* for advertising and discovering services: "What the service does". Each instance of the class *Service* "presents" an instance of the *ServiceProfile* class.
- *Process Model* for describing a service's operation: "How the service works". The class *ServiceModel* captures the property which "describes" a *Service* instance.
- *Grounding* for specifying how to interoperate with a service via messages: "How to access the service" [1]. The *ServiceGrounding* class serves as a "support" property of a *Service* instance.

A service is described by at most one service model, and a grounding must be associated with exactly one service. However, it can be useful for some services to present multiple profiles and/or groundings. An OWL-S Profile describes a service as a function of what organization provides the service, what function the service computes, and what features specify characteristics of the service. The Process Model specifies how to interact with the service. Profile and Process Model hold two different representations of the same service coinciding in input, output, precondition, and effects (IOPEs). The properties described link the Service Profile class with the Service class and Process Model class. Parameters can conveniently be identified with variables in SWRL (Semantic Web Rule Language), the language for expressing OWL Rules. Finally, the Grounding class specifies an unambiguous way of exchanging I/O data: communication protocol, message formats and other service-specific details. An important goal for Semantic Web markup languages, then, is to establish a framework within which these descriptions are made and shared.

OWL-S supports two categories of services; one is *atomic* services where a single web-accessible computer program, sensor, or device is invoked by a request, performs its task and perhaps produces a response to the requester. There is no ongoing interaction between the user and the service. In contrast, *complex* or *composite* services are formed of multiple primitive services and may require an extended interaction/ conversation between the requester and the set of services to be utilized. This distinction suggests us to add to every KP agent an OWL-S specific capability to make easier the interaction of device applications with web services.

While Service Profile and Service Model are abstract representations, the Service Grounding deals with the concrete level of specification. In order to inputs and outputs of a process to be realized concretely as messages, Web Services Description Language (WSDL) carries out the OWL-S grounding mechanism with a consistent binding. In OWL-S a binding is an abstract object with two properties: *toParam* (name of the parameter) and *valueSpecifier* (value description). It results in an effort to provide a specification as concise as possible in variety of situations. WSDL contains document and process information in XML format to bind to a concrete network protocol and message format allowing the service developer to benefit from WSDL's functionality and similar languages (as SOAP, HTTP GET/POST, and MIME) for message exchange. As the OWL-S concept of grounding is generally consistent with WSDL's concept of binding, WSDL is used as the ground of an OWL-S atomic process. For OWL-S/WSDL grounding both languages are needed and complementary. Both languages overlap in the area where WSDL's abstract types, which are used to characterize inputs and outputs of services (specified in XML Schema) correspond to OWL-S' DL-based abstract types OWL classes. However, WSDL/XSD is unable to express the semantics of an OWL class and OWL-S cannot express the binding information that WSDL captures in messages [1]. Therefore, an OWL-S/WSDL grounding uses OWL classes as abstract types of message parts declared in WSDL and then relies on WSDL binding constructs to specify the formatting of the messages.

Table 5.1 Basic elements to define a service.

Basic Elements to define a Service	
OWL-S Element	**Definition**
Type	Data type definition to describe the exchanged messages.
Message	Abstract definition of the data being transmitted.
PortType	Set of abstract operations. Each operation refers to an input message and output messages.
Binding	Concrete protocol and data format specifications for the operations and messages defined by a particular *PortType*.
Port	Address for a binding defining a single communication endpoint.
Service	Used to aggregate a set of related ports.

Grounding OWL-S with WSDL and SOAP (assuming HTTP as transport mechanism) involves the construction of a WSDL service description with all the usual parts (types, message, operation, port type, binding, and service constructs). The essence of an OWL-S/WSDL grounding can be summed up by creating an instance of the OWL-S Grounding class which includes all required information regarding relationships between relevant OWL-S constructs and WSDL constructs [1].

The basic elements for defining a service are described in Table 5.1.

The most basic and concrete class in OWL-S, *WsdlAtomicProcess-Grounding* establishes the grounding mechanism details within a WSDL specification. In order to formalize the details of the grounding, Table 5.2 shows the main properties in the *WsdlAtomicProcessGrounding* class.

5.6.1 Proposed OWL-S Python binding

As OWL-S language is an ontology itself, in order to express preconditions and effects is combined with SWRL (Semantic Web Rule Language), the language for expressing OWL Rules based on OWL DL and Lite and RuleML (Rule Markup Language). Inputs/Outputs are subclasses of SWRL variables. Discovery and composition operate on description logic reasoning.

Table 5.2 Properties representing the OWL-S grounding class *WsdlAtomicProcessGrounding*.

OWL-S Grounding Class	
OWL-S Grounding Property	**Definition**
wsdlVersion	URI indicating WSDL version.
wsdlDocument	URI of WSDL document of the referring grounding.
wsdlOperation	URI of WSDL operation corresponding to the atomic process.
wsdlService and *wsdlPort* (optional)	URI of WSDL service (or port) that offers the given operation.
wsdlInputMessage	Object containing URI of the WSDL message definition carrying inputs of the given atomic process.
wsdlInput	Object containing a list of mapping pairs, (instance of *WsdlInputMessageMap*) with *wsdlMessagePart* property-URI of input object (*owlsParameter*) or *xsltTransformation* property (string or URI) which generates the message part from an instance of the atomic process.
wsdlOutputMessage	Analogue to *wsdlInputMessage*.
wsdlOutput	Analogue to *wsdlInputs*.

Below we show a simple Python example to describe a situation in which the user is watching TV and his PVR (Personal Video Recorder) is On. In order not to miss part of the program and to be able to speak without noise, we would like the PVR to pause automatically when the user receives a phone call. This application could be modeled as Listing 5.1 shows.

```
1 if phone.isRinging(user.getPhoneNo)) & PVR.isOn():
2   PVR.setPause()
```

Listing 5.1 Python Rule Example

A rule expressed in an OO language for modeling a KP behavior can involve different services among its components. We could consider each call to a Python function as a different atomic service which, combined, form a composite service. Once a rule is created or developed, as we have done in

Python, it needs to go through different phases in its execution cycle for it to be completely deployed. The phases start with the installation of the rule in the *Smart Space* until its uninstalation or removal. Next the rule sequential phases are enumerated:

- Rule INSTALLED: All instances of the different KPs involved in the rule must be known and available in the SIB, in other words, a rule is installed when its involved KPs are connected to the *Smart Space* and registered in the SIB.
- Rule RESOLVED: A rule is resolved when for each call to a Python function there exists a specified and available service grounding. Then the rule can be considered as registered in the *Smart Space*. Access control to the KP services is also checked and is here where OWL-S acts as intermediate specification. Without service formalization and grounding in OWL-S the binding of the rule realization would become language dependent. Consequently, OWL-S helps achieving multi device interoperability.
- Rule STARTED or SUBSCRIBED: When the rule's corresponding services have requirements or conditions constraining its triggering, the rule will not become active until they are satisfied. The rule is installed and resolved but is blocked. This means that an asynchronous subscription to the SIB is done in order to get a notification callback when the constraints are met. When that occurs, the rule is released and becomes active or published.
- Rule ACTIVE or PUBLISHED: A rule is active or published when the service has been started, triggered and the execution grounding has been realized. The corresponding effects or updates are published in response to subscriptions.
- Rule UNINSTALLED: The rule is removed from the SIB in the *Smart Space* for it to no longer take effect.

In order to generalize the service binding to Python we have to be conscious that not only *atomic* and *composite* services exist. In applications including simple Python rules as the previous conditional construct (Listing 5.1) we can find *active* and *passive* services. Pasive services translate to queries checking availability in the SIB and subscriptions. Active services can, however, translate Python calls into different effects, either actions or changes in the *Smart Space* or simple data updates. For this reason we have had to specify formally the different phases which the rule passes through during its binding. Here appears also our motivation for extending the existing

KPs interaction in the rule cycle deployment

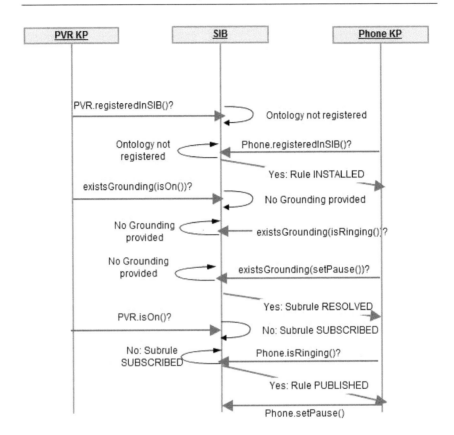

Figure 5.7 Sequence diagram representing the KP's interaction in the rule cycle deployment.

OWL information in the SIB with an OWL-S specification which allows the developer to classify the different types of services and act consequently. For a rule structure to execute in Python, it must be invoked. However, if we structure the OO structures to be represented in form of OWL-S services and introduce them into the SIB, the rules will be always running in the background until their removal is desired.

The Python example in Listing 5.1 could be modeled as the KPs interaction shown in Figure 5.7 representing the rule cycle.

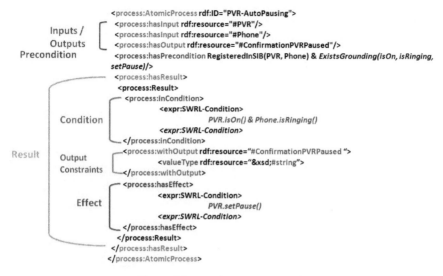

Figure 5.8 Atomic process example.

The previous example showed how one can develop a rule and how can one deploy it in a semantic environment. In order to realize the concrete binding of the service into Python, the OWL-S grounding is needed. We could then model a process as presented and structured in Figure 5.8 and concretize it with an OWL-S grounding as Listing 5.2 specifies.

```
3
4  <!-- The PVR Auto Pausing atomic process -->
5
6  <process:AtomicProcess rdf:ID="PVR-AutoPausing">
7   <process:hasInput>
8    <process:Input ref:ID="PhoneNumber">
9       <process:parameterType rdf:about="&xsd;#string"
          >
10   </process:Input>
11   </process:hasInput>
12   <process:hasInput>
13    <process:Input ref:ID="isPhoneRinging">
14       <process:parameterType rdf:resource="#Phone/
          isRinging">
15    </process:Input>
16   </process:hasInput>
```

```
17    <process:hasInput>
18      <process:Input ref:ID="isPVROn">
19        <process:parameterType rdf:resource="#PVR/isOn"
            >
20      </process:Input>
21    </process:hasInput>
22    <process:hasOutput>
23      <process:Output ref:ID="ConfirmationPVRPaused">
24        <process:parameterType rdf:resource="&xsd;#
            string">
25      </process:Output>
26    </process:hasOutput>
27  </process:AtomicProcess>
28
29
30    <!-- OWL-S Grounding  -->
31  <grounding:WsdlGrounding rdf:ID="SmartPVR">
32    <grounding:hasAtomicProcessGrounding rdf:resource="
          #PVR-AutoPausing"/>
33  </grounding:WsdlGrounding>
34  <grounding:WsdlAtomicProcessGrounding rdf:ID="PVR-
        AutoPausing">
35    <grounding:owlsProcess rdf:resource="#PVR-
          AutoPausing">
36    <grounding:wsdlOperation>
37      <grounding:WsdlOperationRef>
38        <grounding:portType>
39          <xsd:uriReference rdf:value="http://SmartPVR.
              com/PVR-AutoPausing.wsdl#PhonePortType"/>
40        </grounding:portType>
41        <grounding:operation>
42          <xsd:uriReference rdf:value="http://SmartPVR.
              com/PVR-AutoPausing.wsdl#SmartAutoPausing"
              />
43        </grounding:operation>
44      </grounding:WsdlOperationRef>
45    </grounding:wsdlOperation>
46
47    <grounding:wsdlInputMessage rdf:resource="http://
          SmartPVR.com/PVR-AutoPausing.wsdl#
          PausingPVRInput"/>
48    <grounding:wsdlInput>
```

```
49    <grounding:wsdlInputMessageMap>
50      <grounding:owlsParameter rdf:resource="#isPVROn
          ">
51      <grounding:wsdlMessagePart>
52        <xsd:uriReference rdf:value="http://SmartPVR.
            com/PVR-AutoPausing.wsdl#pvr.isOn">
53      </grounding:wsdlMessagePart>
54    </grounding:wsdlInputMessageMap>
55  </grounding:wsdlInput>
56  <grounding:wsdlInput>
57    <grounding:wsdlInputMessageMap>
58      <grounding:owlsParameter rdf:resource="#PhoneNr
          ">
59      <grounding:wsdlMessagePart>
60        <xsd:uriReference rdf:value="http://SmartPVR.
            com/PVR-AutoPausing.wsdl#PhoneNr">
61      </grounding:wsdlMessagePart>
62    </grounding:wsdlInputMessageMap>
63  </grounding:wsdlInput>
64  <grounding:wsdlInput>
65    <grounding:wsdlInputMessageMap>
66      <grounding:owlsParameter rdf:resource="#
          isPhoneRinging">
67      <grounding:wsdlMessagePart>
68        <xsd:uriReference rdf:value="http://SmartPVR.
            com/PVR-AutoPausing.wsdl#phone.isRinging">
69      </grounding:wsdlMessagePart>
70    </grounding:wsdlInputMessageMap>
71  </grounding:wsdlInput>
72
73  <grounding:wsdlOutputMessage rdf:resource="http://
        SmartPVR.com/PVR-AutoPausing.wsdl#PVROutput"/>
74  <grounding:wsdlOutput>
75    <grounding:wsdlOutputMessageMap>
76      <grounding:owlsParameter rdf:resource="#
          ConfirmationPVRPaused">
77      <grounding:wsdlMessagePart>
78        <xsd:uriReference rdf:value="http://SmartPVR.
            com/PVR-AutoPausing.wsdl#pvr.SetPause()">
79      </grounding:wsdlMessagePart>
80    </grounding:wsdlOutputMessageMap>
```

```
81    </grounding:wsdlOutput>
```

Listing 5.2 OWL-S Atomic Process and Grounding Example

5.7 Conclusions and Future Work

A solution for application development was presented in this chapter integrating ontologies and the *Smart-M3* platform. First, developing a tool for mapping OWL to OO languages (available in Python and C) providing complete control over ontologies was described. Second, a middleware was constructed encapsulating the communication with the *Smart Space*. This module allows Python to write applications by using the generated Ontologies APIs. Later a case study implicitly showed the communication among agents through *Smart-M3* Space achieving device interoperability.

For these ontologies to allow composition of services, we finally proposed a deployment through a fixed grounding in the SIB by using an OWL-S description of each subservice. Thus, by extending the previous framework with OWL-S information making it available in the SIB we add support for service interaction and composition.

We conclude by showing the suitability of *Smart Space* for ubiquitous applications where physical environments adapt to the user and the surrounding information is reusable and dynamic. With the integration of these different components with *Smart-M3*, rapid development of context-aware applications for *Smart Space* is made available so that agents can share information independently of which device they are embedded in.

In the future, we aim at creating a context processing library with a Python module for embedding rules expressions [21]. Other challenges to be tackled are e.g. SIB consistency related issues or efficient subscriptions implementation. Then, use cases for other environments can be applied.

Acknowledgment

The presented work was funded through the ICT-SHOCK DIEM project by TEKES (Finnish Funding Agency for Technology and Innovation).

List of Abbreviations

- *Smart-M3*: Multi-domain, Multi-device and Multi-vendor (M3) *Smart Space* platform.

- SIB: Semantic Information Broker.
- KP: Knowledge Processor.
- SSAP: Smart Space Access Protocol.
- OWL: Ontology Web Language.
- OWL-S: Ontology Web Language Services.
- OWL-DL: Ontology Web Language - Description Logic.
- OO Programming: Object Oriented Programming.
- WSDL: Web Service Definition Language.
- WSMO: Web Service Modeling Ontology.
- UPnP: Universal Plug and Play.
- RDF: Resource Description Language.
- SOAP: Simple Object Access Protocol.
- HTTP: Hypertext Transfer Protocol.
- DAML: DARPA Agent Markup Language.
- DARPA: Defense Advanced Research Projects Agency.
- XML: Extensible Markup Language.
- API: Application Programming Interface.
- MIME: Multipurpose Internet Mail Extensions.
- SWRL: Semantic Web Rule Language.
- RuleML: Rule Markup Language.
- WQL: Wilbur Query Language

References

[1] Describing Web Services using OWL-S and WSDL: http://www.daml.org/services/owl-s/1.1/owl-s-wsdl.html.

[2] OWL 2. [online]:http://www.w3.org/tr/owl2-overview/.

[3] OWL-S. [online]:http://www.w3.org/submission/owl-s/.

[4] Smart-M3 Ontology to Python API Generator: http://sourceforge.net/projects/smart-m3/files/smart-m3-ontology_to_python-api_generator_v0.9.1beta.tar.gz/.

[5] Smart-M3 software at sourceforge.net, release 0.9.4beta, May 2010. [Online]. Available: http://sourceforge.net/projects/smart-m3/.

[6] The Protege Ontology editor and knowledge acquisition system: http://protege.stanford.edu/.

[7] WSMO Grounding: http://wsmo.org/tr/d24/d24.2/v0.1/.

[8] Grigoris Antoniou and Frank van Harmelen. A semantic web primer. *International Journal of Ad Hoc and Ubiquitous Computing*, 2:212–223, 2008.

[9] Matthias Baldauf, Schahram Dustdar, and Florian Rosenberg. A survey on context-aware systems. *International Journal of Ad Hoc and Ubiquitous Computing*, 2, 2007.

[10] Dipanjan Chakraborty, Anupam Joshi, Tim Finin, and Yelena Yesha. Service composition for mobile environments. *Mob. Netw. Appl.*, 10:435–451, August 2005.

[11] Harry Chen, Tim Finin, and Anupam Joshi. An ontology for context-aware pervasive computing environments. In *Proceedings of the Workshop on Ontologies in Agent Systems*, 2003.

[12] Oleg Davidyuk, Nikolaos Georgantas, Valérie Issarny, and Jukka Riekki. MEDUSA: Middleware for End-User Composition of Ubiquitous Applications. In *Handbook of Research on Ambient Intelligence and Smart Environments: Trends and Perspectives*. IGI Global, 2010.

[13] Anind K. Dey and Gregory D. Abowd. The Context Toolkit: Aiding the development of context-aware applications. In *Workshop on Software Engineering for Wearable and Pervasive Computing*, 2000.

[14] Jing Dong, Yongtao Sun, Sheng Yang, and Kang Zhang. Dynamic web service composition based on OWL-S. *Science in China Series F: Information Sciences*, 49:843–863, 2006.

[15] Tao Gu, Hung Keng Pung, and Da Qing Zhang. A middleware for building context-aware mobile services. In *Proceedings of IEEE Vehicular Technology Conference (VTC)*, 2004.

[16] Andre Kaustell, M. Mohsin Saleemi, Thomas Rosqvist, Juuso Jokiniemi, Johan Lilius, and Ivan Porres. Framework for Smart Space Application Development. In *Proceedings of the International Workshop on Semantic Interoperability (IWSI 2011)*, 2011.

[17] Shalil Majithia, David W.Walker, and W.A.Gray. Automated web service composition using semantic web technologies. In *Proceedings of the International Conference on Autonomic Computing (ICAC04)*, 2004.

[18] R. Oppermann and M. Specht. A context-sensitive nomadic exhibition guide. In *Proceedings of Second Symposium on Handheld and Ubiquitous Computing*, pages 127–142, Springer, 2000.

[19] W. Prinz. NESSIE: An awareness environment for cooperative settings. In *Proceedings of the Sixth European Conference on Computer-Supported Cooperative Work*, pages 391–410, 1999.

[20] M. Roman, C. Hess, R. Cerqueira, and A. Ranganathan. A middleware infrastructure for active spaces. In *IEEE Pervasive Computing*, 2002.

[21] M. Mohsin Saleemi, Natalia Diaz, Johan Lilius, and Ivan Porres. A framework for context-aware applications for smart spaces. In *Proceedings of ruSMART 2011: The 4th Conference on Smart Spaces*, 2011.

[22] M. Mohsin Saleemi and Johan Lilius. End-user's service composition in ubiquitous computing using Smartspace approach. In *Proceedings of The Sixth International Conference on Internet and Web Applications and Services (ICIW 2011)*, 2011.

[23] Evren Sirin, James Hendler, and Bijan Parsia. Semi-automatic composition of web services using semantic descriptions. In *Proceedings of workshop on Web Services: Modeling, Architecture and Infrastructure (ICEIS2003)*, pages 17–24, 2002.

[24] Z. Song, Y. Labrou, and R. Masuoka. Dynamic service discovery and management in task computing. In *Proceedings of Mobile and Ubiquitous Systems: Networking and Services (Mobiquitous2004)*, 2004.

[25] Espen Suenson, Johan Lilius, and Ivan Porres. OWL web ontology language as a scripting language for Smart Space applications. In *Proceedings of the International Symposium on Rules, RuleML*, 2011.

[26] Mathieu Vallee, Fano Ramparany, and Laurent Vercouter. Flexible composition of smart device services. In *Proceedings of the 2005 International Conference on Pervasive Systems and Computing (PSC-05)*, Las Vegas, pages 27–30, 2005.

[27] Paul Wisner and Dimitris N. Kalofons. A framework for end-user programming of smart homes using mobile devices. In *Proceedings of the Consumer Communications and Networking Conference, CCNC*, 2007.

[28] X.H. Wang, D.Q. Zhang, T. Gu, and H.K. Pung. Ontology based context modeling and reasoning using OWL. In *Workshop Proceedings of the 2nd IEEE Conference on Pervasive Computing and Communications*, 2004.

6

Organization of Quality-Oriented Data Access in Modern Distributed Environments Based on Semantic Interoperability of Services and Systems

Stanisław Polak and Renata Słota

AGH University of Science and Technology, al. A. Mickiewicza 30, 30-059 Krakow, Poland; e-mail: polak@icsr.agh.edu.pl

Abstract

The objective of this chapter is to discuss issues related to the organization of quality-oriented data access in modern distributed environments such as clouds or grids based on semantic interoperability of services and systems. We present the application of semantic technology with the use of ontologies for modeling Mass Storage Systems (MSS) and organization of access to data stored in such systems. We have developed a model of MSS based on which two ontologies have been created. This model and ontologies allow the interoperability of MSS systems and services for monitoring and estimation of MSS performance, which are needed for quality-oriented organization of data access. By using semantics, we managed to raise the level of interoperability of the existing solutions enabling us to address the new challenges related to delivering high quality data storage services. We present the use cases of developed ontologies for the construction of data and storage management systems.

Keywords: mass storage systems, QoS, data management, storage management, data grid, storage cloud, virtual organization, SLA.

S.F. Pileggi and C. Fernandez-Llatas (Eds.), Semantic Interoperability: Issues, Solutions, and Challenges, 131–152.

6.1 Introduction

The IEEE [8] defines interoperability as

> *The ability of two or more systems or components to exchange information and to use the information that has been exchanged*

When the exchanged information should be automatically interpreted by these systems a semantic interoperability is needed. According to the European Interoperability Framework [7]

> *Semantic Interoperability enables systems to combine received information with other information resources and to process it in a meaningful manner.*

Automation of processes, such as, e.g., efficient management of data storage systems, in a changing and heterogeneous environment will require extensive cooperation and understanding between the components of the system. Such understanding and cooperation (semantic interoperability) can provide suitably constructed ontologies.

We present the application of semantic technology with the use of ontologies for modeling heterogeneous Mass Storage Systems (MSS) (see Section 6.1.1) and organization of access to data stored in modern distributed environments such as Clouds or Grids. We show a model of MSS and two ontologies developed on this basis. This model and ontologies allow us to achieve the interoperability of MSS systems and services for monitoring and estimation of MSS performance (see Section 6.1.2), which are needed for quality-oriented organization of data access.

For this reason, in this chapter we show how to create ontologies in order to make our services and systems semantically interoperable what will give us the ability to automate some management processes.

6.1.1 Mass Storage Systems Heterogeneity

In computing, mass storage refers to non-volatile memory, designed for long term storage of large amounts of data, as opposed to RAM and ROM. MSS are systems consisting of storage hardware and computers running storage related software. They differ from each other due to their architecture, functionality and technology [10]. Typical examples of MSS are disk arrays or tape libraries connected to computers equipped with a Hierarchical Storage Management (HSM) system for their management.

Disk array is a storage system which contains a collection of several to several hundred physical disks which are arranged in groups of several to tens of Redundant Array of Independent Disks (RAID).

HSM is a storage system in which the files based on adopted policies are automatically migrated from expensive media (per byte stored) to less expensive storage media. Migrating files are replaced with headers, which from the user's perspective are indistinguishable from the original files. When a user wants to use the file, the data are sent back to the place of the header. The most commonly used policy implies that new files which are often used, are stored on fast and short access time media called caches (e.g. disk arrays), while rarely used files are automatically migrated to the slower and cheaper storage media (e.g. magnetic tapes).

Usage of the above-mentioned systems in the environments where the data access time is limited by Service Level Agreement (SLA) [16] or Quality of Service (QoS) constraints, requires monitoring and prediction of storage performance [1]. This monitoring and prediction are also important in distributed systems with replica management service for data access optimization [20].

6.1.2 Storage Performance Monitoring and Prediction Services

Accurate performance prediction for a given I/O data transfer is a difficult task because there are many elements, such as a large number of MSS parameters, components and software modules which can influence the performance of MSS, and dynamism of the aforementioned environment in which such systems work [19]. Examples of such elements are: system load, caches and data migration policy. In order to predict the performance, the MSS should be equipped with monitoring services and estimation services.

The former services gather 'up to date' information about the MSS state using sensors. The sensors should be implemented for each MSS used in a given environment because of MSS heterogeneity [12]. Estimation services predict the MSS performance for a future data transfer based on the data provided by the monitoring services. The estimation services differ from each other in terms of the type of MSS for which the performance will be predicted and in terms of prediction method (statistical, rule-based, or simulation estimation).

6.1.3 **Problem and Proposed Solution**

The estimation and monitoring services should exchange information and use it without special load, in other words, they must be interoperable. So it is necessary to develop a common model that describes the MSS and their performance parameters. Such a model has been developed – the CIM based Common Mass Storage System Model (C2SM). The idea of exploiting the C2SM model by the monitoring and estimations services for data management with QoS is described later in this chapter.

The representation of the model in the form of an ontology provides new possibilities like, e.g., semantic interoperability of the services. One of the more popular languages to describe data in the form of ontologies is Web Ontology Language (OWL) [14]. In OWL data can be represented in the form of classes or as so-called datatype properties. Depending on the used form of representation, the ontologies cover different areas of application as described later. For this reason, we propose two ontologies: the OntoStor and OntoStor-ATN ontologies. In the OntoStor ontology performance parameters are represented as datatype properties, while in the OntoStor-ATN performance parameters are represented as classes.

Here, first we present the state of the art on modeling MSS, and then the afore-mentioned model of MSS and ontologies describing such systems developed to achieve interoperability of the systems and services are then presented. Finally, we discuss the problem of usage of the developed ontologies, e.g., we present a semantic based QoS monitoring system and a toolkit that facilitates the process of data management for data intensive applications.

6.2 **State of the Art**

Models of informations systems can can be divided into non-semantic and semantic ones. The most popular non-sematic models are: CIM, SMI-S, and GLUE, while the most popular sementic model are: SUMO, OpenCyC, ConceptNet and COSMO. All these models are insufficient to fully describe the MSS systems as well as performance parameters, but there are works in terms of extending these models as described in Section 6.3.1.

6.2.1 **Common Information Model**

Common Information Model (CIM) [2] is an open standard developed by the Distributed Management Task Force (DMTF). It is a hierarchical, object oriented model of management elements in the Information Technology

(IT) environment. CIM is a general model, not bound to a particular implementation. It consists of two parts: an Infrastructure Specification and a Schema.

CIM Infrastructure Specification contains a description of object oriented meta-model based on Unified Modeling Language (UML). The basic elements of this meta-model are, e.g., Classes, Properties, and Methods.

CIM Schema contains a set of predefined classes, their properties, methods, and dependencies among these classes. CIM Schema consists of two separated layers: Core Model and Common Models. The first of these defines the basic dictionary (common classes) for the latter. Common Models define a set of predefines models which are independent of any implementation or technology, and describe particular areas of management.

CIM is the main component of Web-Based Enterprise Management (WB-EM) systems used for a distributed management of computing environments [26].

CIM has many schemas relevant to MSS: Storage Devices, Storage Services, Storage Capabilities and Settings, Storage Statistics, Physical Component.

6.2.2 Storage Management Initiative – Specification

Storage Management Initiative – Specification (SMI-S) [21] is a CIM based standard formulated by Storage Networking Industry Association (SNIA), which defines an interface to manage Storage Area Network (SAN), which is a heterogeneous data storage environment consisting of data storage devices, data storage systems, and management applications. In SMI-S, CIM classes are grouped in profiles (e.g. disk array), and each of them can have subprofiles.

In SMI-S 1.1 which is the ANSI standard, there are four groups of profiles: Storage, Host, Fabric Topology, and Server. The first group relates to the management of data storage devices, while the other ones relate to the management of other elements of SAN, like, e.g., a Fibre Channel Fabric. With all these groups, only the first has profiles which directly regard MSS but they do not represent the needed performance parameters.

6.2.3 Grid Laboratory Uniform Environment

Grid Laboratory Uniform Environment (GLUE) [9] is a conceptual, object-oriented, information model of Grid environments, and its main aim is to provide interoperability among elements of the Grid infrastructure.

The basic elements of the model are main entities, which represent the core concepts of the Grid environment: resource, service, location, etc. Conceptual models of computing and storage services are defined based on the main entities. These entities are described with UML Class Diagrams.

Neither GLUE, CIM nor SMI-S fully represent the MSS systems, e.g., the HSM systems as well as most of the performance related parameter of MSS are not directly represented, so we decided to create our own model of MSS that takes these systems and these parameters into account. Due to the fact that only CIM offers the greatest opportunities to expand, we decided to use this as the basis for our model.

6.2.4 Ontology Based Models of Mass Storage Systems

There are many upper ontologies such as those described briefly below. There are also many domain ontologies like, for example, the QoSOnt ontology described below, which are useful in the process of management of QoS based systems. Because the upper ontologies are too general and too broad, and the QoSOnt is too specific, we decided to build our own ontology by converting the above-mentioned C2SM model using the 'cim2owl' tool [23].

SUMO (Suggested Upper Merged Ontology) [22] is an upper ontology resulting from the integration of publicly available ontologies. It maps all the concepts in the WordNet [27] lexical database. It is composed of about 20,000 terms and about 70,000 axioms. It consists of the Mid-Level Ontology (MILO) and many domain ontologies, like, e.g., ontology of countries and regions or ontology of physical elements.

OpenCyc [13] is an "open source" version of the Cyc – knowledge base and common-sense reasoning engine. It includes all the Cyc terms (non-proprietary) and about a million assertions regarding the terms. The structure of the Cyc knowledge base allows for automated reasoning and drawing conclusions. One of the first practical applications of the system is CycSecure, which examines the actual safety of the network by performing simulations of attacks on the network.

ConceptNet [4] is competitive to OpenCyc knowledge base. Compared to the latter, it is easier to use. Accumulated knowledge and tools are available as

Web services. The data is collected from ordinary people and it is represented in the form of a semantic network.

COSMO (COmmon Semantic MOdel) [3] is an OWL (Web Ontology Language) upper ontology. It is based on OpenCyc and SUMO ontologies. Some elements inherited from the ontologies have significantly different interpretation which reflects a different way of representing some of the basic categories. As a result, despite their similar basic intended meanings they have a slightly different logical properties. The current version offers only the inference rules that provides OWL.

QoSOnt [5] (a QoS Ontology for Service-Centric Systems) is an ontology for QoS. It consists of three layers, and each of them is an ontology. First, the lowest layer is comprised of two ontologies. The first one defines the general concepts related to QoS, while the second defines the concepts concerning units and methods for their conversion. The middle layer includes ontologies that define the different QoS attributes and their metrics. At the highest level are domain-specific ontologies.

6.3 Models of Mass Storage Systems and Their Application for QoS Based Data and Storage Management

The models described in Section 6.2 do not fully address the performance aspects of MSS which are needed for proper management of storage in a distributed environment with QoS constraints, and therefore we created our own model.

6.3.1 The CIM Based Common Mass Storage System Model

The C2SM model describes four kinds of MSS: the HSM systems, disk array systems, local disk systems and the cluster file systems. This model consists of a set of classes (see Figures 6.1 and 6.2) describing the current state of a specific MSS, its configuration (e.g. total capacity of MSS, maximal read transfer rate), and the parameters of physical devices and media (e.g. number of tape tapes, tape block size).

The AGH_StorageSystem class is the main class and it can store common information for all kinds of MSS. The AGH_DiskArraySystem, AGH_LocalDiskSystem, AGH_ClusterFileSystem and AGH_HSMSystem classes represent four kinds of MSS, currently included in the model.

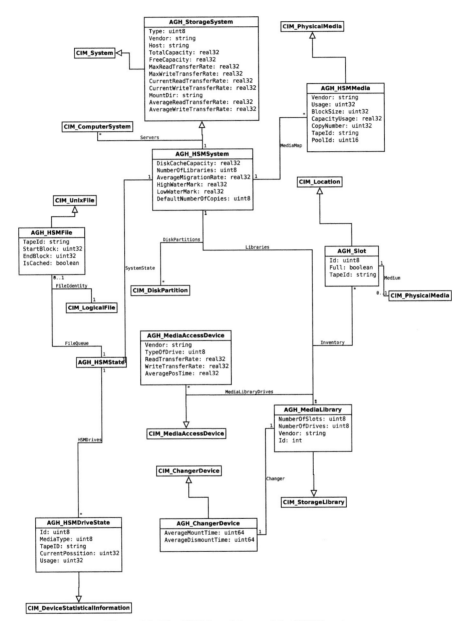

Figure 6.1 The C2SM model – model of HSM systems.

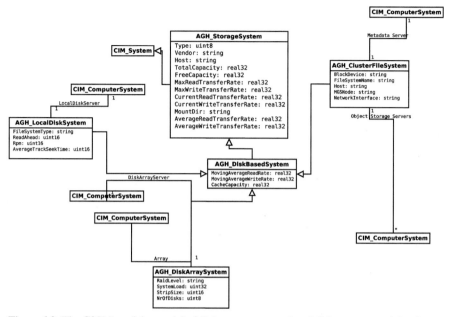

Figure 6.2 The C2SM model – model of disk array systems, local disk systems and the cluster file systems.

The AGH_DiskArraySystem class contains information about disk array systems. These systems consist of a disk array (the CIM_ComputerSystem class) and a server (the CIM_ComputerSystem class).

The AGH_LocalDiskSystem class stores information about local disk systems. These systems comprise a server (the CIM_ComputerSystem class) having hard disks.

The AGH_ClusterFileSystem class describes cluster file systems. Components of this system are:

- server processing and providing metadata (Metadata Server) – the CIM_ComputerSystem class
- servers processing and providing objects (files) (Object Storage Servers) – the CIM_ComputerSystem class

The AGH_HSMSystem contains information common for the HSM systems. These systems include:

- a server (the CIM_ComputerSystem class),
- media libraries (the AGH_MediaLibrary class),
- disks (the CIM_DiskPartition class).

Media libraries (AGH_MediaLibrary) contain:

- slots (the AGH_Slot class),
- a changer device (the AGH_ChangerDevice class),
- drives (the AGH_MediaAccessDevice class).

The AGH_HSMMedia, AGH_HSMDriveState, AGH_HSMFile, and CIM_LogicalFile classes describe the states of media, drives, and files stored in HSM respectively.

The developed model goes beyond the existing models by providing classes for representing MSS (HSM system, disk array, file server) and more performance related properties like data transfer parameters, state of media drives and libraries, queue of pending requests, etc. Such a model allows for more accurate monitoring and provides necessary data to storage system performance estimation services.

6.3.2 Application of the CIM Based Common Mass Storage System Model for QoS Based Data and Storage Management

The above-mentioned model was used in a distributed environment with monitors and estimators for data and storage managing with QoS. The layered view of this environment is shown in Figure 6.3.

Storage Nodes (SN) are complete MSS (like HSM, disk array, file server) which store files and provide access point for accessing physical files. The C2SM model is used by the CIM software which provides data to the MSS monitoring service (MSS Monitor). The aim of the performance estimation services (MSS Estimator) is to provide on request performance prediction result for the given physical storage location. The role of the Metacatalog is to provide a global filename space and directory hierarchy, and mapping logical file names to physical file names. Physical file names (replicas) are accessible by Data Transfer Services. The Client Application has access to their data via Data Access API. SLA Checker monitors the degree of fulfillment of the SLA contract. In the case of data access performance degradation appropriate action is taken, e.g., another file replica is created. All data access requests are tracked by MSS Resource Scheduler. Its role is also to assign storage resources based on prediction results obtained from MSS Estimators. Replication Service is responsible for replica management tasks (like creation, removal, consistency keeping). MSS Resource Scheduler and Replication Service are a part of the Management Layer and they are

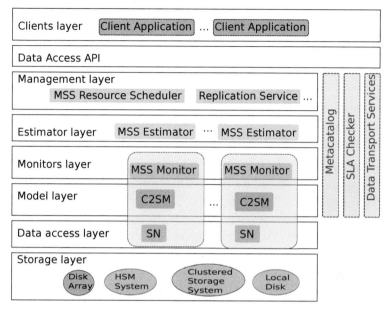

Figure 6.3 Layered view of the distributed storage environment with monitors and estimators for data and storage management with QoS.

examples of services which can be used for data and storage management. When, in this case, estimators and monitors are used, QoS-aware management can be achieved. Using the model layer using the C2SM model enables interoperability between monitors and estimators.

6.4 The OntoStor Ontologies

By using the C2SM model we obtained interoperability between monitoring and estimation services. The next challenge was to achieve semantic interoperability between these services. In order for these systems to communicate with each other and understand each other, an ontology which is a research result of the OntoStor [24] project was created. Thanks to this ontology the afore-mentioned systems:

- can semantically interpret the data which they exchange,
- can retrieve information based on many different criteria,
- have access to data that fit into the context, that is, they have access only to those data which concern them.

The aim of the "OntoStor" research project is to work out an ontology based methodology concerning organization of data access in knowledge supported distributed computational environment with respect to various types of storage systems. The semantic description allows for efficient use of MSS and for easier creating and integration of data and storage access applications.

6.4.1 OntoStor Ontology – MSS Performance Parameters as Datatype Properties

Because a suitable ontology which describes MSS and their performance aspects had not been found, we decided to create such an ontology. Based on the C2SM model described above, the OntoStor ontology was created (see Figure 6.4).

At the beginning, the C2SM model was written in Managed Object Format (MOF), and then converted into Web Ontology Language (OWL) format using the "cim2owl" tool. Next the result file was modified by a human, e.g.: redundant components (individuals, properties, classes) were removed, closure axioms for some classes were added, new properties, and individuals representing existing MSS systems were created. As a result we obtained an ontology in which the classes of the C2SM model were represented by OWL classes, and the class properties by datatype properties in OWL. For example, the 'BlockSize' property of the AGH_HSMMedia class is represented in OWL as shown in Example 1.

```
<owl:DatatypeProperty rdf:about="#AGH_HSMMedia__BlockSize">
    <rdfs:domain rdf:resource="#AGH_HSMMedia"/>
    <rdfs:range rdf:resource="&xsd;unsignedInt"/>
</owl:DatatypeProperty>
```

Example 1: Representation of an example class property of the C2SM model in the OntoStor ontology.

The four kinds of MSS mentioned in Section 6.1 are represented by the following classes: AGH_HSMSystem, AGH_DiskArraySystem, AGH_Cluster-FileSystem and AGH_LocalDiskSystem. Their components are represented by the following classes: CIM_ComputerSystem, AGH_MediaLibrary, AGH_HSMMedia, and CIM_DiskPartition. Components of media libraries of the HSM system are represented by the AGH_ChangerDevice class, the AGH_MediaAccessDevice class, and the AGH_Slot class. The AGH-_HSMMedia, AGH_HSMDriveState, AGH_HSMFile, and CIM_LogicalFile

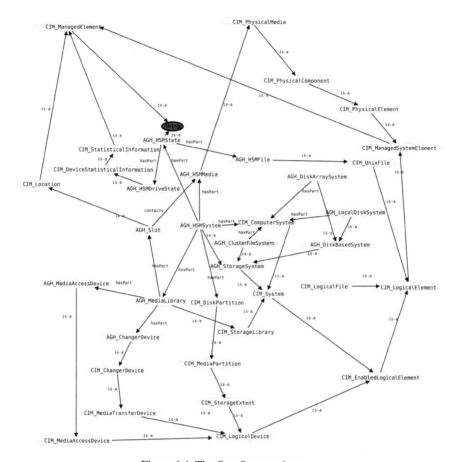

Figure 6.4 The OntoStor ontology.

classes represent information about states of media, drives, and files stored in HSM.

Using this ontology we described the resources in our testing environment, i.e., the HP 660ex magneto-optical library and the ATL 7100 tape library, and their components. In Example 2 we present a fragment of an OWL file containing the description of the above-mentioned magneto-optical library.

As we can see this ontology contains semantically described information about a concrete value of a concrete parameter of MSS. Thanks to this ontology, monitoring and estimation services are able to find a kind of MSS

```
<owl:Thing rdf:about="#Magneto_Optical_library_HP_660ex">
  <rdf:type rdf:resource="#AGH_MediaLibrary"/>
  <AGH_MediaLibrary__NumberOfSlots
    rdf:datatype="&xsd;unsignedByte">
      128
  </AGH_MediaLibrary__NumberOfSlots>
  <AGH_MediaLibrary__NumberOfDrives
    rdf:datatype="&xsd;unsignedByte">
      4
  </AGH_MediaLibrary__NumberOfDrives>
  <AGH_MediaLibrary__VendorString
    rdf:datatype="&xsd;string">
      HP
  </AGH_MediaLibrary__VendorString>
  <hasPart rdf:resource="#Drive_HP_5200"/>
  <hasPart rdf:resource="#Slot_1_HP"/>
  <hasPart rdf:resource="#Slot_2_HP"/>
</owl:Thing>
```

Example 2: Ontological description of the HP 660ex magneto-optical library.

and their components, based on values like numbers or strings, e.g., "find media libraries which have four or less drives". This is needed for example for choosing media library which is appropriate for QoS provisioning in a distributed storage system.

6.4.2 OntoStor-ATN Ontology – MSS Performance Parameters as Classes

The ontology described above does not cover all application areas, e.g., it cannot be used to identify a kind of MSS based on the names of the attributes, therefore an alternative version of the ontology – the OntoStor-ATN (Attribute Name) ontology was created (see Figure 6.5). For simplicity, this ontology defines two main concepts:

AGH_Parameters – basic class for concepts representing resources, i.e., MSS and their components

AGH_Attributes – basic class for concepts representing attributes of resources

Subclasses of AGH_Parameters, like e.g., the AGH_ServerParameters class, describe concepts which can be identified with concrete resources,

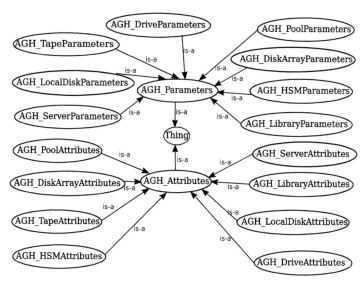

Figure 6.5 The OntoStor-ATN ontology.

e.g., a server. This ontology does not contain individuals of these concepts. Subclasses of the second main class, i.e., AGH_Attributes, define concepts of resource attributes of individual resources, e.g., the AGH_DiskArrayAttributes concept represents attributes of disk arrays. In this case, individuals of the AGH_*Attributes classes are defined in the ontology, e.g., the 'BlockSize' class property of the C2SM model is represent as the 'BlockSize' individual of the OWL AGH_TapeAttributes class. All individuals are associated with concepts of resources by the 'hasValue' restriction – in the case of the afore-mentioned property, it is associated with the AGH_TapeParameters concept as shown in Example 3.

In the same way we assign other individuals. As a result this ontology contains a complete set of concepts related to MSS and individuals representing their attributes.

6.5 Cases of the Use of Ontologies for the Organization of Quality-Oriented Data Access in Modern Distributed Environments

Now the possibilities of building a number of components necessary in modern distributed environments for the organization of quality-oriented data

```
<AGH_TapeAttributes rdf:about="BlockSize">
    <rdf:type rdf:resource="&owl;Thing"/>
    <AGH_Attributes__isDynamic rdf:datatype="&xsd;boolean">
      true
    </AGH_Attributes__isDynamic>
</AGH_TapeAttributes>

<owl:Class rdf:about="AGH_TapeParameters">
  <rdfs:subClassOf rdf:resource="AGH_Parameters"/>
  <rdfs:subClassOf>
    <owl:Restriction>
      <owl:onProperty rdf:resource="#hasAttribute"/>
        <owl:hasValue rdf:resource="BlockSize"/>
      </owl:Restriction>
    </rdfs:subClassOf>
    ...
</owl:Class>
```

Example 3: Representation of an example class property of the C2SM model in the OntoStor-ATN ontology.

access with usage of the above-mentioned ontologies are presented. We show a few use-cases:

- Description of storage resources with the OntoStor ontology
- Monitoring of QoS parameters of storage resources
- Storage management for data intensive application in Grid environments

We describe why in the presented use-cases, specific ontologies must be used, and what the results are of the use of appropriately developed ontologies.

6.5.1 Description of Storage Resources with the OntoStor Ontology

As mentioned previously the OntoStor ontology is used to ensure semantic interoperability of the monitoring and estimation services so that they are able to semantically interpret the data which they exchange, to retrieve information based on many different criteria, and to have access to data that fit into the context. In the ontology, performance parameters are represented as datatype properties.

As was mentioned in Section 6.4.1, we described as examples the HP 660ex magneto-optical library and the ATL 7100 tape library, and their

Table 6.1 The OntoStor ontology – sample queries.

Query	Result – the names of the individuals
Find all devices equipped with the HP 5200 drive `hasPart value Drive_HP_5200`	• Magneto_Optical_library-_HP_660ex
Find all media libraries which have 4 drives `AGH_MediaLibrary__NumberOfDrives value 4`	• Magneto_Optical_library-_HP_660ex • Tape__Library__ATL7100

Table 6.2 Example keywords of Manchester OWL Syntax and their counterparts symbols in the DL.

Manchester OWL Syntax keyword	Description Logic symbol
some	$\exists$
only	$\forall$
value	$\ni$
and	$\sqcap$
or	$\sqcup$

components. The first of these consists of four HP 5200 drives and has 128 slots. The library (see Example 2) is represented by the `Magneto_-Optical_library_HP_660ex` individual of the `AGH_MediaLibrary` class and information about the number of drives and slots is an appropriate data property in the individual. The HP 5200 drive is an individual of the `AGH_MediaAccessDevice` class. The ATL 7100 tape library is represented in the same way. It has 4 drives, 100 slots, and the name of the individual is `Tape_Library_ATL7100`.

In Table 6.1 sample queries to the ontology are shown. The first column contains first the verbal description of the request and then a query expressed in the Manchester OWL syntax [6]. This syntax is used by the Protege-OWL editor when editing class expressions and it is also used by DL Query – a standard Protege-OWL plugin which enables searching a classified ontology. This syntax is based on Description Logic (DL) style syntax and uses English language keywords instead of DL symbols (see Table 6.2).

The results of the queries, i.e., individuals that represent the matching devices, are located in the second column.

We can see that it is possible to find appropriate MSS depending on the requirements.

6.5.2 Monitoring of QoS Parameters of Storage Resources

In this section we discuss the problem of usage of the OntoStor-ATN ontology – we present semantic based QoS monitoring systems based on it.

It is assumed that QoS aspects of storage resources are described by different QoS parameters (metrics). For different kinds of resources, these metrics can have different meaning. Monitoring QoS parameters is a complex issue because it requires taking into account many aspects such as, e.g., (i) mapping high-level QoS parameters for low-level parameters obtained from direct monitoring of resources (see the C2SM model), (ii) distribution and heterogeneity of resources. This complexity demonstrates the need for semantic description of the QoS aspects of storage resources.

In this use-case metrics and methods of metric obtaining for specific types of resources are described by an enriched version of the QoSOnt ontology. In Example 4 definition of a sample metric 'ExampleMetric' is shown. This metric has a general metric type that represents total storage capacity, the metric unit is GB and the expected metric value should be greater than 6.

```
<PLGridCNO:TotalStorageCapacity rdf:about="#ExampleMetric">
  <rdf:type rdf:resource="&owl;Thing"/>
  <QoSOntology:hasValue rdf:datatype="&xsd;int">6</QoSOntology:hasValue>
  <QoSOntology:hasPredicate rdf:resource="&QoSOntology;greater"/>
  <qosont2:hasUnit rdf:resource="&PLGridCNO;gigabytes"/>
</PLGridCNO:TotalStorageCapacity>
```

Example 4: Definition of the sample metric.

A method of metric obtaining defines the specific way of calculating (e.g., sum, mean value) and attributes that are used in these calculations. In Example 5 definition of a sample method ('ExampleMethod') of metric obtaining is presented – this method simply consists of the direct reading of a single attribute, i.e., 'TotalDACapacity' – this attribute is defined in the OntoStor-ATN ontology and represents total capacity of a disk array.

```
<owl:Thing rdf:about="#ExampleMethod">
  <rdf:type rdf:resource="&QoSOntology;SingleParamMethodOfMeasurement"/>
  <QoSOntology:monitoredParameter rdf:resource="&www;TotalDACapacity"/>
</owl:Thing>
```

Example 5: Definition of the sample method of metric obtaining.

The described monitoring system named SMED [15, 18] uses the Pellet OWL reasoner [25]. As a result of use of the reasoner a new inferred ontology model is created. This program allows us to assign the general metrics classes to storage resources individuals, since specific metric classes and specific methods of measurement metrics are applied automatically based on the resource type, e.g., in the above case the 'ExampleMethod' method is automatically assigned to the 'ExampleMetric' metric.

6.5.3 Storage Management for Data Intensive Application in Grid Environments

The enriched version of the OntoStor-ATN ontology is used in the FiVO/QStorMan toolkit [17] and its task is to facilitate the process of data management for data intensive applications. This toolkit is a part of the FiVO framework [11].

Data intensive applications require access to resources that offer high-speed data access and high reliability. The aforementioned toolkit specifies a location of storage nodes, corresponding to previously specified non-functional requirements for running data intensive applications. The toolkit consists of several components, for example (i) a distributed knowledge base called GOM containing information about the application requirements and semantic descriptions of storage resources, (ii) a system SMED monitoring storage resources and providing variety of information about QoS parameters (see Section 6.5.2), and (iii) a system searching storage nodes which meet the QoS requirements – the SES system.

All information, i.e., information about non-functional requirements and information about available storage resources, is stored in the form of an ontology. The latter is described by the enriched version of the OntoStor-ATN ontology. Compared to the original, this ontology has been updated with several new concepts, like, e.g., physical storage resources and virtual storage resources, i.e., resources which are distributed, consist of several physical devices but are seen as a single resource. The ontology is used by the SMED and SES systems in order to find and select an appropriate method of measuring the parameters of the resource for the type of resource, and to find (for the given non-functional requirements) the most appropriate storage node. Implemented QStorMan tool for the PL-Grid infrastructure accelerated the data-intensive operation of 20 to 40%.

6.6 Conclusions

In this chapter we propose a model of MSS – the C2SM model. It is used in distributed environment using monitoring and estimation services for data managing with QoS. With this model we achieved interoperability between monitoring and estimation services, and interoperability with other services located in the upper layers of data and storage management system.

It is not easy to manage MSS with QoS in changing environment, e.g, MSS systems can be expanded or can change, definitions of metrics can be changed, there may be some different definition of QoS parameters for different types of MSS. Models such as C2SM cope well with changes in a heterogeneous environment MSS systems, enabling the monitoring and estimation in a changing environment without changing the services, but the construction of the next level of data and storage management systems in modern distributed environments requires reliance on semantics. Hence, here we suggests the construction of the OntoStor and OntoStor-ATN ontology and we have shown the possibility of their use in the several use-cases. The reason for creating the latter ontology was that the OntoStor ontology does not cover all applications areas, i.e., it allows one to find a kind of MSS and their components, based on values like numbers or strings but it cannot be used to identify a kind of MSS based on the names of the attributes. As a result it was impossible to obtain semantic interoperability of components of systems like QoS storage monitoring or storage management for data intensive applications in Grid.

Acknowledgments

This research is partially supported by the MNiSW grant No. N N516 405535 and AGH-UST grant No. 11.11.120.865. We would like to thank Professor Jacek Kitowski for scientific assistance during the research and Dr. Darin Nikolow for fruitful cooperation.

References

[1] Rajkumar Buyya, Chee Shin Yeo, Srikumar Venugopal, James Broberg, and Ivona Brandic. Cloud computing and emerging IT platforms: Vision, hype, and reality for delivering computing as the 5th utility. *Future Generation Computer Systems*, 25(6):599–616, 2009.

[2] Common Information Model (CIM) Standards. http://www.dmtf.org/standards/cim/, last access 16 Feb, 2012.

[3] COmmon Semantic MOdel (COSMO). http://micra.com/COSMO/, last accessed 16 February 2012.

[4] ConceptNet. http://csc.media.mit.edu/conceptnet, last accessed 16 February 2012.

[5] Glen Dobson, Russell Lock, and Ian Sommerville. Qosont: An ontology for QoS in service-centric systems. In *Proceedings of the Conference on e-Science All Hands Meeting*, 2005.

[6] Nick Drummond. The Manchester OWL syntax. http://www.co-ode.org/resources/reference/manchester_syntax/, last accessed 16 February 2012.

[7] Semantic Interoperability Centre Europe. https://joinup.ec.europa.eu/asset/page/practice_aids/what-semantic-interoperability, last accessed 16 February 2012.

[8] Anne Geraci. *IEEE Standard Computer Dictionary: Compilation of IEEE Standard Computer Glossaries*. IEEE Press, Piscataway, NJ, 1991.

[9] Grid Laboratory Uniform Environment (GLUE). https://forge.ogf.org/sf/projects/glue-wg, last accessed 16 February 2012.

[10] Krishna Kant. Data center evolution. *Computer Networks*, 53(17):2939–2965, 2009.

[11] Bartosz Kryza, Łukasz Dutka, Renata Słota, and Jacek Kitowski. Security focused dynamic virtual organizations in the grid based on contracts. In Paul Cunningham and Miriam Cunningham (Eds.), *Collaboration and the Knowledge Economy, Issues, Applications, Case Studies*, volume 5, II, pages 1153–1160. IOS Press, 2008.

[12] Darin Nikolow, Renata Słota, and Jacek Kitowski. Storage QoS aspects in distributed virtualized environment. In *Proceedings of CLOUD COMPUTING 2010, the First International Conference on Cloud Computing, GRIDs, and Virtualization*, Lisbon, Portugal, November 21–26, pages 110–115. IARIA, 2010.

[13] OpenCyc. http://www.opencyc.org/, last accessed 16 February 2012.

[14] OWL Web Ontology Language Overview. http://www.w3.org/TR/owl-features/, last accessed 16 February 2012.

[15] Kornel Skałkowski, Joanna Sendor, Renata Słota, and Jacek Kitowski. Application of the ESB architecture for distributed monitoring of the SLA requirements. In *Proceedings of 2010 Ninth International Symposium on Parallel and Distributed Computing (ISPDC2010)*, Istanbul, July 7–9, 2010.

[16] Łukasz Skitał, Maciej Janusz, Renata Słota, and Jacek Kitowski. Service level agreement metrics for real-time application on the grid. In Roman Wyrzykowski, Jack Dongarra, Konrad Karczewski, and Jerzy Wasniewski (Eds.), *Proceedings of 7th International Conference (PPAM2007)*, Gdansk, Poland, September, Lecture Notes in Computer Science, Vol. 4967, pages 798–806. Springer, 2007.

[17] Renata Słota, Dariusz Król, Kornel Skałkowski, Michał Orzechowski, Darin Nikolow, Bartosz Kryza, Michał Wrzeszcz, and Jacek Kitowski. A toolkit for storage QoS provisioning for data-intensive applications. *Computer Science*, 13(1), 2012.

[18] Renata Słota, Darin Nikolow, Marcin Jarząb, Kornel Skałkowski, and Jacek Kitowski. High performance data access aspects in national data storage. In *Proceedings of Cracow Grid Workshop – CGW'08*, Kraków, October 13–15, pages 335–340. ACC-Cyfronet AGH, Kraków, 2009.

[19] Renata Słota, Darin Nikolow, Stanisław Polak, Marcin Kuta, Marek Kapanowski, Kornel Skałkowski, Marek Pogoda, and Jacek Kitowski. Prediction and load balancing system

for distributed storage. *Scalable Computing Practice and Experience, Special Issue: Grid and Cloud Computing and Their Application*, 11(2):121–130, 2010.

[20] Renata Słota, Darin Nikolow, Łukasz Skitał, and Jacek Kitowski. Implementation of replication methods in the grid environment. In Peters M.A. Sloot, Alfons G. Hoekstra, Thierry Priol, Alexander Reinefeld, and Marian Bubak (Eds.), *Proceedings of Advances in Grid Computing, European Grid Conference (EGC2005)*, Amsterdam, The Netherlands, February 14–16, Lecture Notes in Computer Science, Vol. 3470, pages 474–484. Springer, 2005.

[21] Storage Management Initiative Specification (SMI-S). `http://www.snia.org/tech_activities/standards/curr_standards/smi`, last accessed 16 February 2012.

[22] Suggested Upper Merged Ontology (SUMO). `http://www.ontologyportal.org/`, last accessed 16 February 2012.

[23] The CIM2OWL tool. `http://fivo.cyfronet.pl/trac/fivo/wiki/KIG/CIM2OWL`, last accessed 16 February 2012.

[24] The OntoStor project. `http://www.icsr.agh.edu.pl/ontostor/`, last accessed 16 February 2012.

[25] The Pellet OWL reasoner. `http://clarkparsia.com/pellet/`, last accessed 16 February 2012.

[26] Web-Based Enterprise Management (WBEM). `http://www.dmtf.org/standards/wbem`, last accessed 16 February 2012.

[27] WordNet. A lexical database for English. `http://wordnet.princeton.edu/`, last accessed 16 February 2012.

Subject Index

About the Editors

Salvatore Flavio Pileggi received his M.Sc. in Computer Engineering from the University of Calabria (Italy) in 2005 and his Ph.D. in Communications from the Polytechnic University of Valencia (Spain) in 2011. After a short experience as Software Engineer, he worked as researcher in the Communications Department at the Polytechnic University of Valencia (Spain). He is currently researcher in Health & Wellbeing Technologies Group in ITACA institute at Polytechnic University of Valencia (Spain). He is actively working on several open research issues such as semantic technologies, distributed computing, sensor networks, eHealth and eInclusion.

He is author/co-author of more than 30 scientific papers that appear in some of the most relevant international journals, magazines and conference proceedings. His first publication about the integration of WSN with Grid Computing (2006) received a Best Paper Award. Since 2010 he is chairing the international Workshop on Semantic Sensor Web (SSW) and, since 2011, the International Workshop on Semantic Interoperability (IWSI). For the last years, he has been involved in several Spanish Government founded research initiatives as well as in a significant number of European Commission founded research projects.

Carlos Fernandez-Llatas is M.Sc. in Computer Engineering (2001) from the Universidad Politecnica de Valencia (Spain), and he received the PhD degree in the Pattern Recognition and artificial Intelligence Program of the same university in 2009. He is currently foresight and research coordinator in Health & Wellbeing Technologies Group in ITACA Institute at Universidad Politecnica de Valencia. He has more than 10 years of experience in research and participated in more than 20 projects through IV, V VI and VII European Framework Program and Spanish Goverment funded projects. He has published over 40 scientific papers.

155

His research is mainly focused on process management learning, representation and execution techniques and its automatic learning for their applications in health and human behaviour modelling. He also has a wide experience on software platforms and Business Process Management technology.